W9-DIR-492

THE FATAL CONFRONTATION

Wilbur R. Jacobs (Photograph by Robert Schlosser.)

THE FATAL CONFRONTATION

Historical Studies of American Indians, Environment, and Historians

Wilbur R. Jacobs

INTRODUCTION BY ALBERT L. HURTADO

Published in cooperation with the
University of New Mexico Center for the American West

University of New Mexico Press
Albuquerque

HISTORIANS OF THE FRONTIER AND AMERICAN WEST
RICHARD W. ETULAIN, SERIES EDITOR

For Alex DeConde, friend, tennis partner, and colleague

First Edition

Library of Congress Cataloging-in-Publication Data

Jacobs, Wilbur R.
The fatal confrontation : historical studies of American Indians, environment, and historians / Wilbur R. Jacobs ; introduction by Albert L. Hurtado. — 1st ed.
p. cm. — (Historians of the frontier and American West)
"Published in cooperation with the University of New Mexico Center for the American West."
Includes bibliographical references.
ISBN 0-8263-1764-2 (cloth)
1. Frontier and pioneer life—United States—Historiography. 2. Indians of North America—Historiography. 3. Man—Influence of environment—United States—Historiography. 4. Man—Influence on nature—United States—Historiography. 5. United States—Historical geography—Historiography.
I. Title. II. Series.
E179.5.J25 1996
973'.072—dc20 96-4499
CIP

CONTENTS

List of Illustrations

PREFACE

Some fifty years ago, when I was a doctoral student under Louis Knott Koontz at the University of California, Los Angeles, I began my study of Native Americans by reading geography, anthropology, and history. In this body of early scholarship, I saw clearly what was missing in our traditional accounts of Indian history: They told no real concrete and sophisticated story of the Native Americans and their culture.

William C. Macleod's *American Indian Frontier* introduced me to the works of the famed ethnologist, Frank G. Speck. Like one of my favorites Clark Wissler, Speck observed major environmental themes in Native American ways of life. Out of such intellectual encounters, reinforced by my contacts with other prominent writers like William Brandon, Carl O. Sauer, Sherburne F. Cook, Bernard DeVoto, William Fenton, Robert Heizer, Henry F. Dobyns, Francis Jennings, and Allan Nevins, my own publications multiplied and expanded into the historiographical and ethnoenvironmental fields. In my youthful career I stood under the intellectual umbrellas of Francis Parkman and Frederick Jackson Turner, but as the following essays demonstrate, I soon stepped out into the rain to become a critic of these writers. Whatever value my essays might have, may perhaps be attributed to their representing assessments of ethnoenvironmental history in the recent past.

These essays are republished in their original form with the exception of a few minor textual corrections. I wish to express my gratitude to the journals that first edited and published my work and the readers who initially read and engaged my interpretations. I am especially grateful to my former students and colleagues and to the staffs of the Huntington

Library and the Library of the University of California, Santa Barbara. Jean Bonheim, Martin Ridge, Ann Firor Scott, John Steadman, Alan Jutzi, Timothy Braatz, Alexander DeConde, Calvin Martin, and Paul Zall have also placed me in their debt.

My heartfelt gratitude goes out to Albert Hurtado, who studied under me at the University of California, Santa Barbara, who has enlightened the scholarship of Native Americans in the Spanish Borderlands, and who graciously agreed to introduce this volume. I also want to thank Richard W. Etulain, director of the University of New Mexico Center for the American West, for inviting me to submit and publish a selection of my work in his edited series, Historians of the Frontier and American West. I owe a special debt to University of New Mexico Press editor Durwood Ball, who contributed expert commentary to this project.

For permission to reprint my essays, I am grateful to *The Journal of American History, The Pacific Historical Review, The American Historical Review, The Western Historical Quarterly, American West Magazine, The William and Mary Quarterly, and The American Historical Association Newsletter*.

Wilbur R. Jacobs
Huntington Library
San Marino, California

INTRODUCTION

Albert L. Hurtado

Wilbur R. Jacobs was a leading historian of the American frontier when I arrived at the University of California, Santa Barbara, in 1974. I had read some of his revisionist essays and books, but when I first entered his office as a beginning doctoral student, I had no idea of the reception that I would get. I was pleasantly impressed. The friendly Jacobs was genuinely glad to meet me and soon put me at ease. He also got me a job at the campus bookstore within five minutes, a surprising development because I had not expected him to be an employment agent. After calling the bookstore, Jacobs said, "Turn around. Put your hands on the wooden file cabinet." This odd request puzzled me, but I complied. "That," he explained, "was Frederick Jackson Turner's file cabinet." Jacobs customarily gave this initiation to many of his students, and I was suitably awed.

A closer look around Jacobs's office showed that it was a room where someone was hard at work. Papers, letters, and memos littered his desk. There was a typewriter that always seemed to have a half-finished letter scrolled in it. There were two walls lined with books, including a long shelf holding Francis Parkman's works, which were stuffed with notes nearly bursting the spines. "I've read them rather carefully," Jacobs drily explained. There were usually three or four students asking questions of the professor, who was as accessible to the green freshman as he was to the seasoned graduate student. Lively conversations in his office about current events, the latest historical interpretations, and the profession were part of the Santa Barbara scene that I knew in the 1970s.

Jacobs presided over a two-quarter research seminar that annually attracted a dozen or more students with widely varying interests. These seemingly disparate seminar subjects more or less reflected Jacobs's intellectual concerns that had emerged as his career unfolded. Most of Jacobs's graduate students studied with him because he was a leading authority on American Indian history, but others worked in colonial history, the environment, the frontier, and the West. In three years I heard papers that spanned the sixteenth and twentieth centuries and that discussed the Caribbean, Canada, Connecticut, California, and much in between. The seminars usually met in Jacobs's ranch house up in the canyon behind mission Santa Barbara. From the gracious patio we looked down the canyon and out to the Santa Barbara Channel where the sun set in a spectacular splash of color. In that incomparable California landscape we read and debated while Jacobs questioned us and offered suggestions for improvement.

Jacobs's home reflected his personality and interests. He furnished it mainly with colonial pieces, including the Parkman chair. Really, we asked? Well, he explained, it was said to have come from the Parkman family. Anyhow, it was a good example of the Windsor style, and it came with an interesting story. In the living room there was a fireplace with two portraits of Benjamin Franklin on the mantle—a double-dose that symbolized Jacobs's interest in Franklin and his embrace of the philosopher's advice about hard work and perseverance. In another building he had an office that housed his working library, desk, and typewriter. There, with essentially the same view as from the patio, he studied and wrote.

At first glance the interests of Jacobs and his students may seem eclectic, but they represent the rational growth of Jacobs's thought on the history of the American frontier. For Jacobs this intellectual odyssey began during his youth in Altadena, California. Jacobs's father introduced him to University of California, Los Angeles (UCLA) professor Louis Knott Koontz, a specialist in colonial and frontier history, from whom the elder Jacobs had taken extension courses. Koontz was an occasional guest at the Jacobs family's dinner table, which became a forum for the discussion of Frederick Jackson Turner and frontier subjects, especially when other visitors like the noted colonial historian, Lawrence Henry Gipson, came to dine and talk. But it was Koontz who shaped the young Jacobs's thinking more than anyone else. The distinguished professor even took Jacobs on his first visit to the Huntington Library, where the budding scholar would research most of his publications in the following decades. Jacobs's youthful interests matured into a conviction that history was his calling. Eventually, he earned the doctorate at UCLA under the tutelage of Professor Koontz.[1]

While these formative experiences inspired a lifelong interest in Turner and the frontier, Jacobs did not become one of the many uncritical disciples of the master. Koontz had urged Jacobs to write his dissertation on the colonial Indian frontier, and this galvanizing experience changed Jacobs's thinking about the frontier forever. His pioneering dissertation became a book, *Diplomacy and Indian Gifts: Anglo-French Rivalry along the Ohio and Northwest Frontiers, 1748–1763*, a classic ethnohistorical study that explains Indian-white relations in terms of Indian culture and perspectives.[2]

As many other ethnohistorians have learned, Jacobs found that the study of Indian history from an Indian perspective was a transforming process. After he completed his research on the British colonial frontier, he could no longer view Indians as barriers to civilization or as uncouth savages who had no rights worth respecting. Instead, Jacobs showed that Indians had admirable qualities, stern morality, and a just cause—the preservation of their societies and homelands.[3] This revelation led Jacobs to reinterpret American history as he researched and wrote about the frontier. The essays republished here illustrate the breadth of Jacobs's scholarship, which has evolved from several consistent themes related to Indian and frontier history. The essays in this book are grouped in three sections: "The American Frontier Environment," "Contact and Its Consequences," and "Great Interpreters of the Frontier Experience."

Jacobs's conviction that historians have misrepresented Indians has strengthened over time and is a consistent theme in his essays. The articles in "Contact and Its Consequences" demonstrate how he has called on historians to reconsider the European conquest of the western hemisphere in the face of new data and fresh interpretive viewpoints. He sees the recalculation of Indian populations as a key to a new understanding of hemispheric history. If we accept the general proposition that there were more Indians than Europeans in 1492, Jacobs contends, then we must come to a new understanding of the nature of European colonization. And this understanding extends beyond the boundaries of the western hemisphere, as his comparative essay on New Guinea and North America amply demonstrates. As we follow the logic of

Jacobs's writings, we see that Indian history is not merely a localized event prefatory to white settlement but an important and ongoing part of world history.

Jacobs's early work on the people of the eastern woodlands inspired his ever-widening interest in the meaning of the frontier experience. As his career developed, he analyzed the intellectual foundations of frontier and western history, especially the seminal work of Francis Parkman and Frederick Jackson Turner. His interest in these two historians flowed naturally from his own historical concerns. Parkman was America's greatest narrative historian. His multivolume study of the French and British struggle for North America and his classic travel account, *The Oregon Trail*, have influenced historians and lay readers for more than a century. But Jacobs was interested in more than merely enshrining the great literary figure. He sought to illuminate the historical context and intellectual underpinnings of Parkman's work as well as his hefty influence on subsequent generations of historians. Thus, in addition to writing articles and a book about Parkman, Jacobs published a critical edition of Parkman's correspondence.[4]

Jacobs's work on Parkman has continued for more than three decades, and it is not surprising that Jacobs's opinions about the master historian have changed during that time. While Jacobs admires Parkman's novelistic style and recognizes his contribution to the American romantic tradition, Jacobs criticizes him for producing national heroic symbols—symbols that necessarily cast Indians in a negative, brooding light—in books about wilderness struggle.[5] On the other hand, as the selected essays here show, Jacobs has remained open to finding new possibilities in Parkman's character and writings. For example, he discovers in Parkman's love of nature a latent environmentalism that influenced his historical writings. Jacobs also finds in Parkman's work flashes of geographic determinism that later captivated Turner and convinced him that successive encounters with the wilderness shaped American history.

Likewise, Jacobs has been at pains to analyze Turner's contribution to American history. As he did with Parkman, Jacobs spent decades delving into Turner's private papers. This research resulted in edited volumes of Turner's unpublished essays and correspondence that are basic to understanding the development of frontier theorist's thought.[6] Jacobs also published important interpretive essays that have shaped scholars' ideas about Turner's work. Even-handedness characterizes Jacobs's critique of Turner in the selections presented in this volume. He recognizes Turner as a great and influential teacher who shaped the historical profession in its formative years. Yet, he finds that ethnocentrism marked Turner's work and has stained the contributions of his uncritical followers. Although Turner portrayed himself as a dispassionate and scientific scholar seeking truth, Jacobs sees that Turner was wedded to his own ideas and defended them at all costs. Moreover, the pernicious aspects of Turnerian thought continue to resonate in frontier and western history, and Jacobs argues that historians must reinterpret the past in light of information and ideas that have surfaced since Turner's time.

Jacobs's deep study of Parkman and Turner have influenced his scholarship in many ways. His immersion in their classic works has nurtured the direct and open style that characterizes his writings. Jacobs's fine writing style was not derived merely from some mysterious osmotic process. Like all good writers, he has worked hard at his craft. He urged his doctoral students to write *ten* pages a day, but none of them kept up that pace. Jacobs also had practical advice for writer's block. He once counseled me to choose a work by a great stylist like Samuel Eliot Morison and to read a few pages aloud. Then return to my writing, he said, and the words will flow. It is good advice, and Jacobs practices what he preaches. He has published more than one hundred articles and a dozen books, an outpouring of scholarship that contributed important revisions to American history.

One of Jacobs's most important legacies to the historical discipline is his study of the environmental past. This work, too, grew out of his study of American Indian and frontier history. While studying Indian culture, Jacobs encountered the pioneering work of Frank Speck, who argued that Indians were ecologists. This idea intrigued Jacobs, who became increasingly interested in exploring human–land relationships. Jacobs's environmental consciousness had earlier wellsprings, too. His mother was a crusading antivivisectionist, and her convictions about the ethics of animal exploitation re-enforced her son's growing environmental concern. Eventually, Jacobs became an outspoken advocate for animal rights and a vegetarian.

In 1969 a stunning crisis sharpened Jacobs's focus on environmental issues. Several large platforms stood in the Santa Barbara Channel where they quietly extracted petroleum from beneath the ocean floor. One January day, as the roughnecks pulled up a drillstem on one of the platforms six miles offshore, a well blew out. The leaking well released more than 250 million gallons of crude oil, creating a foul-smelling slick that stained twenty miles of beaches and stretched forty miles out to sea. The sticky, viscous substance suffocated and blinded thousands of seabirds and killed some California gray whales as they migrated toward Alaska. The befouling of the channel, one of the most beautiful places in the United States, made Santa Barbara a byword for industrial pollution.

The disaster had happened on the front stoop of the University of California's Santa Barbara campus, and the tragedy deeply affected the whole university community. The spill was especially meaningful for Jacobs and other faculty with environmental interests. Jacobs became even more convinced of the connection between environmental despoliation and the dispossession of the American Indians. Once the environmental basis of Indian culture was destroyed, he reasoned, Indians were easy prey for those who coveted their lands. Thus, Jacobs's selected environmental writings argue that the extinction and near extinction of various animal and plant species, Indian population decline, Euroamerican conquest, the marginalization of Indian tribes, and historians' celebration of these events as "progress" are linked. From the sixteenth century to the present, a long history of brutal frontier conquest and environmental exploitation has built a legacy of extermination. Jacobs does not mince words about his conception of the frontier movement. In his mind, it was a series of holocausts. For Jacobs, this was the dark side of the triumphant Turnerian march of American civilization, the truly savage story hidden behind the Parkman narrative.

In the 1990s many of Jacobs's ideas are in keeping with what some have called the "new western history," although he cheerfully criticizes the work of these younger scholars.[7] When Jacobs first expressed these ideas, however, some historians—especially those who cherished an unreconstructed Turnerian perspective—considered them to be novel revisions, or something worse. Jacobs was among the leaders of the movement, beginning in the 1960s, to make American history more inclusive of diverse ethnic groups. "You must see history from the Indians' point of view," he used to tell his seminars. He insisted that the nation's historical record must include its mistakes as well as its triumphs. In succeeding decades many historians have come to agree with Jacobs's conception of American history. Thus, the essays reprinted here are not merely a sketch of one professor's intellectual growth, but a record of the reconceptualization of American history.

It is difficult to summarize a historian like Wilbur R. Jacobs who has labored so hard for so long. It seems to me that above all he is a historian with a conscience. He believes that he has an obligation to challenge conventional historical interpretations so that history includes all its human actors, human crimes, and miscalculations as well as achievements. Jacobs has a name for this inclusive, conscience-driven historical view: he calls it telling the truth. Some historians believe that they cannot resurrect "truth" from historical sources that are limited, lost, or biased. They think that Jacobs's simple demand is unrealistic, or perhaps quaintly naive. But there are others, including this author, who find his forthright summons to be truly refreshing and inspiring. The following essays will amply illustrate Jacobs's unflinching view of American history. They still confront historians with hard truths and goad them to reconsider comfortable assumptions. All things considered, Jacobs's work is likely to remain provocative and important for some time to come.

NOTES

1. Wilbur R. Jacobs, *On Turner's Trail: 100 Years of Writing Western History* (Lawrence: University Press of Kansas, 1994), xi–xii. I am also fortunate to have several letters from Jacobs (13 October 1994, 26 Octob er 1994, 27 October 1994, and [n.d.] November 1994) that contain biographical data. Unless otherwise stated, information on Jacobs's life is based on these sources and the many conversations that I have shared with him.

2. Stanford University Press published *Diplomacy and Indian Gifts* in 1950. Sixteen years later, the University of Nebraska Press reissued the volume in the Bison Books series as *Wilderness Politics and Indian Gifts: The Northern Colonial Frontier, 1748–1763.*

3. Wilbur R. Jacobs, ed., *Dispossessing the American Indian: Indians and Whites on the Colonial Frontier* (1972; reprint, University of Oklahoma Press, 1985).

4. Wilbur R. Jacobs, ed., *Letters of Francis Parkman,* 2 vols. (Norman: University of Oklahoma Press, 1960).

5. Wilbur R. Jacobs, *Francis Parkman, Historian as Hero: The Formative Years* (Austin: University of Texas Press, 1991).

6. Wilbur R. Jacobs, ed., *Frederick Jackson Turner's Legacy: Unpublished Writings in American History* (San Marino, Calif.: Huntington Library Press, 1965); Jacobs, *The Historical World of Frederick Jackson Turner, with Selections from His Correspondence* (New Haven: Yale University Press, 1968).

7. For his ideas on recent western scholarship, see Jacobs, *On Turner's Trail,* 203–36.

PART I

THE AMERICAN FRONTIER ENVIRONMENT

CHAPTER ONE

FRONTIERSMEN, FUR TRADERS, AND OTHER VARMINTS

An Ecological Appraisal of the Frontier in American History*

> "In writing their histories of this country they have so hastily disposed of this refuse of humanity [the Indians] which littered and defiled the shore and the interior. It frequently happens that the historian, though he professes more humanity than the trapper, mountain man, or gold digger, who shoots one as a wild beast, really exhibits and practices a similar inhumanity to him, wielding a pen instead of a rifle."
>
> THE JOURNAL OF HENRY D. THOREAU.

At the University of California, Santa Barbara, History Department Wilbur Jacobs and his colleague, Roderick Nash, helped shape and lead debates on the environment in American history. Jacobs applied the concept of environment to re-envision American history, literally from the ground up. In this piece, Jacobs criticized triumphalist history that celebrates the settling and civilizing of the Great American Wilderness by farmers, fur traders, cattlemen, and other frontier "heroes." In his clear, concise style, Jacobs deflates these traditional icons of the frontier movement. He recalls their destruction of North American timberlands, mountains, grasslands, rivers, and animal populations and explains the consequences for wildlife, Native Americans, and settlers.

We read in recent news reports that our fish and wildlife are being slaughtered by the thousands. Conservationists warn us that many types of birds and animals are in danger of extinction here and abroad as an expanding human population relentlessly squeezes them out of wilderness sanctuaries. Even breathing is difficult in air dangerously polluted

*From AHA Newsletter 7 (November 1970), 5–11.
Reprinted by permission of the American Historical Association.

by millions of automobiles and a huge industrial complex. Many of our lakes and rivers are contaminated, and even our beaches are blackened by sludge washed in from the sea.

The destruction of our natural environment is usually viewed as a great modern problem, the implication being that only in the twentieth century has the onslaught taken place. There is growing realization, however, that from the beginning of our history we Americans have been both destructive and wasteful of natural resources. It is actually the scale of the damage instead of its newness which forces us, though still reluctantly, to confront the problem today.

We historians must bear some responsibility for the lateness of our awakening, for we really have not done our homework. We have avoided and in most cases ignored the complicated series of historical phenomena that brought about our dilemma. Our histories, particularly our frontier-sectional or "western" histories, tend to give us a glowing get-rich-quick chronicle of the conquest of the continent. As Frederick Jackson Turner wrote in his influential essay of 1893, ". . . the Indian trade pioneered the way for civilization . . . the trails widened into roads, and the roads into turnpikes, and these in turn were transformed into railroads. . . . In this progress from savage conditions lie topics for the evolutionist." The Turnerian theme of "progress" of American civilization has generally reflected itself in American social attitudes toward the wilderness. The Indian is viewed as a "consolidating influence" on frontiersmen who banded together for defense. When the tribesman is brought into the story, he is depicted as a kind of obstacle to the westward movement. The Indian's respect for animal life and reverence for the land, when mentioned, are usually dismissed as superstition. Unlike the Indian, the white man, with a Judeo-Christian ethic stressing man's dominance over nature, had no religious scruples about exploiting the wilderness. From the beginning, fur traders, who had rum to encourage warriors to hunt, were often frustrated by the reluctance of natives to busy themselves in the useful activity of scouring the woods for furs and skins. Modern American social attitudes toward wild animals show a persistence of the fur trader's point of view. Unless a species can be fitted into a category of being particularly "useful" in a commercial sense, there is public apathy about its survival. A good illustration is public acceptance of an extensive government poisoning project designed to exterminate coyotes in several western states, but so indiscriminately and carelessly administered that serious damage is being done to so-called "non-target species" as well.

This strictly utilitarian attitude toward wilderness life, though widespread in American society, has been partially balanced by a countertheme of appreciation of the wilderness by such writers as Francis Parkman, Henry David Thoreau, John Muir, and Aldo Leopold. They identified wild country and wild animals with genuine human freedom. If we have no free-roaming wild life in wild country, they argued, then we eliminate space for that remaining wild thing, the irrepressible human spirit.

Do historians have an obligation to help counteract harmful social attitudes about the environment that run contrary to the best interests of the nation at large? It is a question that has plagued the consciences of some of our best writers, including Francis Parkman and Frederick Jackson Turner. Certainly we historians have no responsibility for what has happened in the past, but we do have access to historic records and knowledge of what earlier generations did or failed to do. The public and students can expect, therefore, that we will make available the fruits of our investigations in a form undistorted by patriotism, prejudice, sentiment, ignorance, or lopsided research. But such a presentation of the American past is, in certain areas of history, not always the rule. This criticism can be applied particularly to specific subject in "frontier" or "westward movement" history.

Historians of the American frontier, for instance, have failed to impress their readers with the utterly destructive impact that the fur trade had upon the North American continent and the American Indian. There are no investigations of the role the fur men had in killing off certain types of wildlife, which in turn had a permanent effect upon the land and upon native and white societies. The traders and their followers, the fur trading companies, are usually depicted as benefactors of the development of American civilization as it moved westward from the Appalachians to the Pacific Coast. Indeed, the story of the fur trade is almost always (and perhaps unconsciously) told with a capitalistic bias. The historian usually expressed a businessman's outlook in describing the development and expansion of this mercantile enterprise. If the fur trade contributed to the rapid economic growth of the country, and it unquestionably did (Walter Prescott Webb argued that it helped to develop a boom economy in the first two centuries of our history), then the implication is the fur trade was a good thing for all Americans. Free furs and skins, free land, free minerals; it was all part of the great westward trek and the "The Development of American Society," according to Turner and his followers. The self-made man, the heroic figure, who conquered the wilderness was the free trapper, the mountain man. Because the history of trading does not naturally attract the reader's interest, historians of the frontier have often gilded their flawed lily with a bit of spurious romanticism. The bear and bison hunter becomes the courageous tamer of the wilderness.

The real question of interpretation here is: who is the real varmint? The bear, or the trapper who killed him? Aside from the fact that bears are sometimes noted for antisocial behavior, our frontier historians have not had a problem in answering such a question because their interpretations have been conditioned by a society steeped in a *laissez-faire* business ideology. Our view of progress—one which permeates all groups of society and leads us to accept without question the need for an expanding economy—is that progress consists in exploitation and growth, which in turn depends on commercialization and the conquest of nature. In our histories we have treated the land more as a commodity than as a resource. We have here in a nutshell the conquistador mentality that has so long dominated the writing of much American history.

Until recently historians have largely ignored the ecological challenge. Yet truthful, interesting American history, double-barreled and difficult to write, is in part the revolting story of how we managed to commercialize all that we could harness and control with our technical skills. It is, in its unvarnished state, an unpleasant narrative of the reckless exploitation of minerals, waterways, soil, timber, wildlife, wilderness, and Indians, a part of a larger man's rape of nature over centuries. Can historians ignore the fact that today we can no longer afford such cruelty and wastefulness? Do we have an obligation to help our fellow Americans in learning the art of frugality in husbanding our resources? If we as historians hope to foster an awareness of the dangers of further greedy exploitation of the land, we must examine our history in detail to see what has led up to our present situation.

The earliest exploiters of the American continent, the first to tamper seriously with the ecological balance that had existed for centuries relatively undisturbed by aboriginal tribes (such as the Iroquois who depended upon an economy closely governed by the ecology of the north-eastern forest), were the colonial traders and trappers. Their successors, the American trapper-frontiersmen, have been blown up into heroes in our histories. In his 1893 essay Turner helped to shape the myth. His enthusiasm for pioneer types proved to be so infectious that we have tended to accept his interpretations, ignoring the fact that men of the trapper-trader fraternity were often unscrupulous, lawless, and hungry for personal gain. These rascals and pleasure-hunters destroyed without scruple countless beavers, otters, large and small carnivores, and the great bison herds that once lived on our grasslands and woodlands. Even Turner in later years reached the conclusion that pioneers were "wasteful and seeking quick results rather than conservation and permanence."

Can we stand back as historians and look at the American western migration as a huge page in social history to see how clearly the story of the frontier advance is also the story of the looting and misuse of the land? The traders who led the procession of pioneers through the Cumberland Gap and South Pass were the vanguard of those who slaughtered beaver, buffalo, and antelope and thus reduced the western Indian tribes to a state of semistarvation, making them easy victims for sporadic white military campaigns. Ironically the individual fur traders, miners, and cattle raisers were, in many cases, ruined by powerful combines devoted to large-scale commercialization of the natural resources of the West. Because of our exaggerated respect for the entrepreneur (or the pioneer or frontiersman, as we have often called him), we as historians have failed to condemn this early rape of the land, just as today—and for the same reasons—we have not joined visibly in the condemnation of industrial pollution.

How can historians assess the damage brought about by allowing commercial interests to override our best national interests? We might begin by an attempt to gauge the effects of the substitution within a half-century of hundreds of thousands of horses and domestic cattle for wild hoofed animals that existed in the huge area of the Louisiana Purchase. What effect did this have on the fertility of the land and the balance of nature? Can we as historians join scientists in calling attention to the important principle that the earth's productivity depends upon an organic cycle, an order of nature in which organic material taken from the earth must be returned to it? The violation of this principle by Americans in the eighteenth and nineteenth centuries was responsible for the destruction of a great virgin wilderness. In this destruction the substitution of annual grasses for the life-sustaining primordial prairie sod of middle America is one of the more momentous happenings in American history, but the subject is something less than a favorite with historians of the frontier.

Those who hope to write about significant historic events of this kind (and consequences arising out of them) will need a sort of knowledge not ordinarily possessed by historians. To study the impact the fur trade had upon America, for instance, we must have more than a beginning acquaintance with ethnology, botany, silviculture, biology, zoology, and indeed much of the physical sciences. Our doctoral programs for budding American historians are in drastic need of revision if we hope to train qualified candidates who can write intelligently about the history of the exploitation of the land. In addition, there is a sophisticated body of thought about the value of wilderness to Americans that should be studied. There is often difference between what wilderness is and what we *think* it is. Francis Parkman, perhaps our greatest historian, trained himself to write about the Indian, the American forest, and early American civilization. At one time a professor of horticulture, he understood and loved the American forest; he understood, therefore, the respect for nature among the Indians. He is a rare example of a historian who took the trouble to educate himself so that he could "manage," as he said, "to tell things as they really happened."

But the writers who followed Parkman did not follow his methods or his example. Those historians who dominated later frontier historical writing were in many respects prejudiced against the land, the wilderness, and the animal life in it. The historiography of the fur trade is a kind of case study of the larger problem of rewriting the history of the American frontier. The old guard of the fur trade history—Hiram M. Chittenden, Harold A. Innis, Douglas MacKay, Reuben Gold Thwaites, Wayne E. Stevens, Frederick Jackson Turner (and some of his pupils and disciples, especially Louise P. Kellogg, Albert Volwiler, and Robert Glass Cleland)—have left us with a one-sided view of the fur trade in American history. Though Chittenden, Turner, and Cleland loved the wilderness and spoke against the despoliation of nature, all bear a responsibility for the romanticizing of the traders. Cleland's approach to the mountain trappers as a "reckless breed of men" conferred on them a spurious glamour as well as a scholarly stamp of approval. It is in the mountain man of the Rockies in the heyday of the fur trade that we found an archetypal hero emerging, an adventurous figure celebrated in virtually every nationalist account of western man's contact with the undeveloped world.

This particular coloration of the mountain man prevails today in influential monographs and textbooks and indicates that perhaps it is time to make a revaluation of such "pathfinders" as George Croghan, Manuel Lisa, Jedediah Smith, William Ashley, James Ohio Pattie, David Jackson, and the Sublette brothers, who were among the first to begin civilization's war against wildlife on the American continent. If an overdue revaluation of these men fits them into the classification of some of the "varmints" they cleaned out, then there are others who might be reclassified for recognition among the heroes of American history. We might even want to take a second look at Daniel Boone, Kit Carson, and Jim Bridger, or the great entrepreneur, John Jacob Astor.

It can be argued that we cannot morally condemn the pioneers for exploitation. They acted in a manner consistent with their circumstances within their concepts of territorial rights, justice, and morality. When the sky was darkened by thousands of pigeons, the normal, expected reaction was to kill them off wastefully. What we can blame is the continuation of such attitudes into an era of scarcity. We should understand our pioneers, perhaps rather than blaming them for what they did.

It can be further argued that the pioneers, who were quite as mercenary as the leaders of large companies, did not understand the long-range consequences of what they were doing to the land. There were few individuals among the farmers, the hydraulic and strip miners, the loggers, or the sheep and cattle men—for that matter, the fur companies and the railroad tycoons—who had any real conception of the vital importance of the resources they were destroying. They did not grasp the significance of muddy streams cleaned of beaver, waterways that had once run deep and clear. Nor did they appreciate the importance of the vital prairie grasses that we plowed under in a few decades. They were often unaware that precious minerals quickly and forever disappeared from our streams and mountains. Indeed, what nineteenth century conservationists talked about the railroads destroying the ecological balance of the Great Plains?

Yet it is not entirely true that the pioneers, the miners, the fur men, and the western entrepreneurs were ignorant of the consequences of their acts. The pioneer's question, "Why should I look after my descendants?"

and his answer, "They ain't done nothin' for me," go back generations in American history. Our ancestors were often intentionally and ruthlessly destructive. In California, for example, the manner and thoroughness in which California's wildlife and groups of aboriginal people were killed is a blackwash on all Americans. The Spaniards and Mexicans of California seemingly were able to live with wildlife without destroying it. But the forty-niners of California's golden age and their followers were wildly wasteful of elk, antelope, bighorns, bears, small fur-bearing animals, grouse, geese, and shorebirds. Thousands of dollars were made in selling game meat to miners. In California alone a great faunal shift took place in the years 1850–1910, only duplicated by prehistoric postglacial terminations of certain species. The California grizzly was pursued until none at all survive today. Here we have a historical example of the dismal story of mass slaughter of wildlife. Alaska faces a somewhat similar crisis today.

The attempt to preserve the wilderness resources left to us surely deserves the support of historians. The old fur trade history was lopsided—unsympathetic to the land and its environment, glorifying the hunter and trapper. It should be rejected by the coming critical generation of American historians. To a degree, the same kind of criticism can be leveled at conventional "frontier" expansion histories of the cattlemen, mining, lumbering, agriculture, business, transportation, industry, and Indians. The assessment will certainly be made; and the sooner it begins, the better. The tunneled frontiersman vision of our past that has engendered a conquistador attitude toward the land will be revised. Our history will surely have new heroes and a new category of destructive varmints. Some historians writing on conservation, Indian history, and related topics have already pointed the way.

The movement to halt the destruction and pollution of the natural environment is now taking root all over the world. We know mankind has now the potential for self-destruction, for irretrievable pollution of the environment as well as for overcrowding the land. American historians either as individuals or working within the framework of our academic organizations (the American Historical Association and its Pacific Coast Branch, the Organization of American Historians, the Western History Association, and related societies) can contribute to the success of the movement to bring about a better balance between man and nature by zeroing in on the origins of the destruction of our environment.

CHAPTER TWO

THE GREAT DESPOLIATION

Environmental Themes in American Frontier History*

Seven years after "Frontiersmen, Fur Traders, and Other Varmints" appeared, Wilbur Jacobs elaborated on environmental themes before members of the Pacific Coast Branch of the American Historical Society. Reaching back to the sixteenth century, his presidential address, "The Great Despoliation: Environmental Themes in American Frontier History," forges a direct link between the destruction of Native American societies and the despoliation of the natural environment. This pathbreaking article, cited repeatedly since its appearance in 1977, shows how Euroamericans transferred their attitudes toward American Indians to the wilderness, making no distinction between the two. North American frontiersmen ruthlessly uprooted and cast aside whoever impeded the progress of "civilization" and furiously exploited the West's natural resources.

I believe environmental themes deserve more attention in American history than they have hitherto received. Environmental history can be a window to a clearer image of the past and can offer us unique perspectives on generally accepted historical concepts of unlimited

Reprinted from Pacific Historical Review *47 (February 1978), 1–26.*

growth, frontier expansionism, and the rapid use of nonrenewable natural resources.[1]

The environmental theme goes back beyond discovery, but my analysis begins with the environmental impact of the European discovery of America, or, as it is now called, "the Columbian exchange."[2] Even at this time we have eyewitness accounts of environmental impacts. The evidence reveals that Americans and their colonial ancestors altered their natural surroundings and set in motion physical and biological processes that have had reverberating effects on the environment. Our environmental past, then, can be viewed as a history of our modification of the earth and the cumulative effects of our actions that have set new natural forces in motion.[3] With this in mind, what were the significant results of drastic changes Europeans and Americans made in the natural world of the Western Hemisphere?

Although the Columbian exchange had profound political implications, its environmental impact was also drastic. When Columbus landed on an island in the Bahamas, which he named San Salvador, the gentle Arawak Indians welcomed him warmly. They were a peaceful people who made pottery, wove cloth, and carried on a farming-fishing, handicraft lifestyle that held little immediate interest for the Admiral because they mined no gold. The Spaniards, however, were impressed with the physical beauty of the native people and the verdure of the island, "the very green trees," and the variety of delicious fruits.[4] Within little more than a century, by the time Jamestown was settled, San Salvador had experienced an environmental transformation. The Arawaks were gone; in their place were Spanish planters who had transformed the land into cotton fields worked by African slaves. By cutting down tropical vegetation and turning fertile land into one-crop agricultural fields, Spanish planters leached the soil of its nutrients. In time, swells of windblown sand filled the interior areas. Eventually the island, with its human, animal, and plant life destroyed, became virtually a desert.

Essentially the same story was repeated later on the larger island, Hispaniola. Here, likewise, the island's fertility and the large native population of skillful farmers (estimated to have developed the most productive fields in the world—cassava, beans, maize, and other crops) had been destroyed by the 1580s.[5] This destruction took place less than a century after the Admiral had made the island his headquarters, divided the land among his followers, and forced Indians to work mines and fields in the land they had once called their home. Indeed, the Indians' mortality was so incredible that it called forth protests of the Spanish Dominican, Bartolomé de las Casas, who tells us that in 1552 there were only a hundred Hispaniola Indian survivors. He goes on to report that island after island had been despoiled, as he says, "totally unpeopled and destroyed." Again and again Las Casas writes that other islands were "waste and desolate," "wholly desert."[6]

Large islands like Cuba and Jamaica, Las Casas wrote, were subjected to the assault. Though Indian mortality on these islands was very high, and though tropical vegetation was altered, nature was able to fight back in later centuries. And, of course, the mainland of North America, the eastern edge of a huge shaggy continent, also felt the impact of the European invasion. The immense land mass at first gave resistance, but everywhere there was depopulation among Indian people after first contacts with Europeans.

Modern scholars now argue that Las Casas's estimates of Indian mortality were modest[7] and that there were as many as 8 million Indians on Hispaniola in 1492, some 25 million in central Mexico, and as many as 10 million people in North America, north of the Rio Grande.[8] And in the whole Western Hemisphere there were probably 100 million, possibly a larger population than in western Europe. All of these Indian people had developed subsistence patterns to support skyrocketing populations through domestication of food plants and through developing skills to harvest wild plants. They were also excellent hunters and fishermen, but they had little resistance to the white man's microbes.

The disease frontier that swept across the Western Hemisphere came with the Europeans, many of them immune carriers. Smallpox, transported by an airborne virus, was undoubtedly the worst Indian killer since it returned repeatedly in epidemic waves to smother survivors of previous attacks. Other diseases, many of which we now call childhood illnesses, such as measles and chickenpox, ravaged whole Indian societies.[9]

We can round the circle of the entire Western Hemisphere, appraising this first great environmental impact, though we are still just beginning to learn of its profound ramifications for the New World. One fact we must now cope with is that it caused a demographic disaster without known parallel. Although scholars may argue whether 50 million or 100 million Indians lived in the New World in 1492, they agree (except for a few stubborn Hispanics)[10] that the new data drastically change our view of what happened in early American history. We can no longer write about a peaceful occupation by English peoples of a fertile Appalachian frontier. Rather we have to deal with the displacement of millions of people, an invasion of Europeans into densely populated lands. And we must understand that the process decimated many Indians who had developed life-styles to support millions of people without obliteration of the land and its resources.

After having studied a mass of evidence in the biological, physical, and social sciences, I am convinced that Indians were indeed conservators. They were America's first ecologists. Through their burning practices, through their patterns of subsistence (by growing, for instance, beans and corn together to preserve the richness of the soil), by creating various hunting preserves for beaver and other animals, and by developing special religious attitudes, Indians preserved a wilderness ecological balance wheel.[11] Even the intensive farming of the Iroquois, without chemical fertilizers and pesticides, protected the ecology of the northeastern forest.[12]

Victor Shelford, in his excellent book on the ecology of North America, argues convincingly that prehistoric America was divided into a number of distinct biotic provinces. In each, Indians, as well as plants, animals, insects, and other forms of life, were integral parts of an ecological niche. Modern Americans, Shelford maintains, have altered or destroyed ninety-eight percent of these original North American ecosystems.[13] Indian people, on the other hand, had lived within them for centuries by developing a land ethic tuned to the carrying capacity of each ecozone. Indians today know these facts, though they are couched in a different kind of language, handed down through centuries by oral recall. As one of the most recent Indian spokesmen, Vine Deloria, Jr., a Sioux, recently wrote, "The land-use philosophy of Indians is so utterly simple that it seems stupid to repeat it: man must live with other forms of life on the land and not destroy it."[14]

But let us return to the subject of historical environmental impacts. There are other environmental changes in our history that illustrate my thesis that Europeans and, later, Americans set in motion new forces which altered the face of the land. I refer, for instance, to the tidal wave of settlement that in one generation occupied most of the territories of the Louisiana Purchase and substituted domestic hoofed animals (cattle, horses, hogs, sheep, and goats) for millions of wild hoofed animals (bison, deer of various kinds, including elk, moose, antelope, and wild sheep and goats). One authority estimates that 100 million wild hoofed animals originally occupied North America, and certainly a large portion of that number were part of the organic ecological circle of the Louisiana land.[15] This environmental event, which swiftly transformed the natural world of a vast area, is hardly mentioned in our general histories, save an occasional word of indignation over indiscriminate slaughter of plains buffalo. The very magnitude of this environmental change, including the dispossession of the Indians, boggles the mind. Scholars like James C. Malin spent a lifetime evaluating some of the ramifications. He finally specialized in the history of plains grasses, many of which evolved from European origins. He and other investigators found hardy descendants of wild buffalo grasses growing along railroad rightaways.[16] Malin's environmental history researches have not had the recognition they deserve.[17]

To detect the themes and patterns of this environmental shift, we can turn to eyewitnesses who tell us much about how and why changes persisted. Early Virginia accounts help us to understand the gradual environmental modifications that took place in the colonial South. In particular, Robert Beverley, early eighteenth-century planter, gives us clues. The soil was so rich, Beverley tells us, that all kinds of crops flourished, but tobacco was favored from the beginning because "it promised the most immediate gain," which, in turn, caused planters to "overstock the market." His fellow Virginians, Beverley complained, "spunge upon the Blessings of a warm Sun and a fruitful soil . . . gathering in the Bounties of the Earth."[18] What Beverley so clearly observed, the extraordinary exploitation of the soil by southern planters, was echoed in the writings of a series of eminent Virginians, including George Washington, Patrick Henry, and Thomas Jefferson. Washington complained that his fields were running into gullies.[19] Henry is credited with the statement, "He is the greatest patriot, *who stops the most gullies*."[20] Though the planters experimented with clover, contour planting, and other devices, soil erosion and depletion persisted. This subject is a constant theme in American agricultural history.[21]

While the farming practices in the middle colonies placed less emphasis upon cash crops which exhausted the soil, there is historical evidence of wasteful consumption everywhere in these colonies. Peter Kalm, a Swedish scientist touring this part of colonial America in the 1750s, was surprised at the destructive habit of settlers who cleared the land, used it for crops, then for pasture, and later moved on to repeat the process in new land. Another traveler observed that in New Jersey in the 1790s there was "stupid indifference" to the land. The Americans, he said, "in order to save themselves the work of shaking or pulling off the nuts, they find it simpler to cut the tree and gather the nuts from it, as it lies on the ground."[22] Forests were regarded as "troublesome growths." A European visitor was astounded to see his landlord casually cut down "thirty two young cedars to make a hog pen."[23]

Another side of the penetration of the wilderness is given by Cadwallader Colden, colonial New York scientist and early ethnohistorian. According to Colden and other sources collaborating his statements, the entire beaver population of what is now the state of New York, then the Iroquois country, was exterminated in the 1640s as a result of the Anglo-Dutch fur traders operating out of Albany.[24] Only some thirty years after Henry Hudson had explored the wilderness river that bears his name, beaver, one of the most beneficial animals in the North American ecological balance, had become a victim of the assault on wildlife. The attack on beaver was persistent and far-reaching, culminating in its extermination in many areas of the Far West. Fortunately, in recent years, this remarkable animal has reappeared in certain locations.[25]

In colonial New England there are reliable accounts of Puritan assaults upon forests, wildlife, and the soil. Carried on almost like the wars against the Indians, the war against the land resulted in cutting down the big trees, killing much of the furbearing animal population, and exhausting the light cover of topsoil.[26] As one midwestern critic alleged, New Englanders were much like their soil, intensely cultivated, but only six inches deep;[27] they left a trail of abandoned farms and old stone fences enclosing former fields and pastures.

Despite this midwestern criticism, the evidence shows that each of the New England colonies, almost from the very beginning, did have an environmental awareness evident in statutes providing for protection of natural resources in the immediate neighborhood of settlements. Statutes, for instance, restricted the unlimited range of livestock, especially hogs, which invaded common pastures and cornfields.[28] Streams were protected from overfishing, forests from overcutting, all of which was a part of a scheme for social control that governed the life-style of entire colonies, even including rebellious Rhode Island.[29] Although some regulations were designed to prevent nearby forests from being cut so that inhabitants would not have to go great distances for firewood, we can discern here, I believe, a respect for the land. This is undoubtedly a factor that has helped to preserve much of the charm that still remains in New England today.

Jeremy Belknap, writing in 1792 of his beloved New Hampshire, goes so far as to tell us that the good life was living in harmony with nature, though he stresses a man-controlled environment. As he phrased it, "Were I to form a picture of a happy society, it would be a town consisting of a due mixture of hills, valleys, and streams of water. The land well fenced and cultivated. . . . " In his good land he sees the need for a good inn, a "practical preacher," a schoolmaster, and a "musical society," but he stresses that the society should be mostly "husbandmen." When all these elements are combined in a beautiful natural setting, one finds, he writes, a situation "most favorable to social happiness of any which the world can afford."[30]

Thomas Jefferson, of course, developed a similar agrarian theme in his *Notes on Virginia*. While he took great pride in the American environment, celebrating its vastness and the superior nature of its denizens to those found elsewhere (noting, for example, "the reindeer [of the Old World] could walk under the belly of our moose"), he believed that the land should not be left to the Indians.[31] Benefits should come to the white farmers who subdue the land. He, indeed, called for immediate population growth, predicting, as did his friend Benjamin Franklin, that the American population would grow exponentially. "The present desire of America," he wrote, "is to produce rapid population by as great importation of foreigners as possible." He went so far as to assert that "our rapid multiplication will . . . cover the whole northern, if not the southern continent."[32]

In Belknap as well as in Jefferson we see clearly a pride in the American landscape as the nation expands and utilizes the bounties of nature. From the birth of the republic, then, there was an ambivalence about appreciating and protecting nature or exploiting the land. Increasingly the evidence of eyewitnesses is that there was environmental distress, but this was seen as part of the penalty of progress as America moved west to occupy the wilderness.

Henry Schoolcraft, explorer, Indian agent, and author of a valuable survey report in 1819, is one of the many government reporters on western lands who details the environmental distress. He visited the Midwest and the Ozark Highlands during the period of early penetration of wilderness resulting from a lead-mining boom. His particularized account of forty-five mines in Burton Township, Missouri, described a mining frontier that left abandoned pits everywhere in its wake as miners moved from place to place in an eager search for riches. "Unwilling to be disappointed," Schoolcraft wrote, "they fall to work and tear up the whole surrounding country." The savagery of the miners in churning up the land was almost paralleled by another group of pioneers, a society of fur hunters whom the explorer called "savage Europeans" because of their crude lifestyle and their wanton destruction of wildlife.[33]

By the time government surveyor David Owen visited the Midwest several decades later, in 1852, swarms of emigrants from eastern American and European hives had occupied the land. Mining and fur trapping pioneers were no longer of great importance, but Owen had much to say about the assault of the lumbermen. Writing mainly on the geology of the Midwest, Owen does, however, pause to record what we now know was the disappearance of the great Wisconsin pine forests. On the Wisconsin River, alone, he tells us, there were twenty-four mills running forty-five saws cutting 19½ million board feet of lumber and 3 million board feet of shingles. The annual value of the lumber for the millowners was $123,000.[34] Mills on the other nearby rivers did four and five times as much business. Why the tremendous demand for wood? Owen says that there was an "immense consumption of building material" in cities already built. Moreover, there were numerous new towns and villages "which spring up," he says, "as if by magic, along the shores of the Mississippi and its tributaries."[35] Owen's comments help us to understand the forces behind the rapid plundering of America's midwestern timber resources.

During this westward surge of the 1850s, and in the decades immediately following, there were other government surveys, including Clarence

King's *Geographical Exploration of the Fortieth Parallel;* the surveys led by Ferdinand V. Hayden after the Civil War, which were instrumental in the creation of Yellowstone National Park; and George M. Wheeler's *Report upon . . . Surveys West of the One Hundredth Meridian* of 1889.[36] They are collectively a mine of data on environmental history. The earlier Zebulon Montgomery Pike report of 1810 and the Stephen H. Long account of 1821 are excellent background reading for understanding the origins of the so-called myth of the "great American desert" of the Southwest.[37]

The Pike and Long commentaries help us to appreciate the most valuable of all the post-Civil War accounts, John Wesley Powell's *Report on the Lands of the Arid Region of the United States.*[38] This single work, acclaimed by our best scholars of conservation history, was the first study of the land to call for a scientific and environmental understanding of the West and its wilderness resources. Powell, a former professor, a one-armed veteran of the Civil War, and an explorer of the Colorado River, was concerned about the American abuse of arid lands. His *Report* of 1878 was at once a treatise and a sermon on the need to revise our western land policies, which were based upon farming in the humid areas east of the Mississippi. He argued that we needed detailed maps of the arid country that identified mineral, timber, pasturage, and irrigable lands. Along with this he called for a series of land laws based on new concepts and providing for regional water planning. Eighty acres, for instance, should be sold to an irrigation farmer, but 2,500 acres were the minimum needed for a grazing operation.[39]

But Powell was battling ignorance, greed, and disinterest in conservation. In fighting speeches and published articles he elaborated on his report and also confronted a backlash of arguments. One of the strange, unscientific myths that Powell combated was the idea that "the rain follows the plow." All a farmer had to do was to plant trees in an arid area to change the climate.[40]

Powell's forceful statement on environmental abuse was but one of many eyewitness accounts that collectively helped bring about new policies and land laws. While Powell was laboring for reform as a government geologist, a contemporary of his, George Perkins Marsh, provided the first overall appraisal of what had happened to the land in America and in the world at large by the middle of the nineteenth century. Marsh, a brilliant self-made ecologist, a one-time school teacher, lawyer, businessman, and diplomat, first saw the great despoliation in miniature in his native state of Vermont where the Green Mountains, even faster than the tobacco lands of the South, had lost their topsoil.[41] After fires, timbermen, insects, and wheat farmers had done their worst, Vermonters turned loose their herds of sheep to denude the slopes. The resulting erosion, Marsh said, caused "deserts in summer and seas in autumn and spring."

In his book, *Man and Nature, or Physical Geography as Modified by Human Action*, Marsh showed that man's impact upon nature had become so powerful that nature could no longer heal herself. Encyclopedic in his research, probing sources in languages of antiquity as well as those of modern times, he described land abuse in the world from the time of the pollution of Rome to the 1860s. Knowledgeable in the physical and biological sciences, in engineering, in the law, and in methods of historical research, he demonstrated that the cumulative effect of all the plows and axes was ecological catastrophe in the world. He traced the violent attacks through aqueducts, reservoirs, canals, dams, diking, flooding, and the needless killing of birds, insects, reptiles, and mammals.[42] His essay, for instance, on the value of sand dunes is strikingly modern. His comment that extensive mining might alter the magnetic and electric condition of the earth's crust is still ahead of its time.[43] As Stuart Udall has noted, Marsh's book, *Man and Nature*, was the beginning of land wisdom in this country.[44] What Marsh told us—and all other writers have had to build upon his observation—was that in nature there had been ecological balance until the arrival of man brought a serious imbalance.[45]

While Marsh's and Powell's observations were materializing into books that would, in time, help bring about an ecological consciousness in America, there were many others extolling the virtues of expansionism, land exploitation, and corralling the Indians on reservations. The most representative, and perhaps the most persuasive, of all these writers was the economist, statistician, former Indian superintendent, and one-time president of the Massachusetts Institute of Technology, Francis A. Walker.[46]

It was Walker, also a superintendent of the U.S. Census in 1870, 1880, and 1890, whose views eloquently illustrated the prevailing ideas of the day, which were based upon the social Darwinism of Herbert Spencer and other writers. Extolling the kind of men who rose "from stage to stage in intellectual, moral, and physical power,"[47] Walker at the same time praised the "genius" of Americans who had a "fire of Americanism," inventiveness, adaptability, and and ability to meet and mingle in the vast western territories, which "had no history of their own." In tracing the United States' center of population as it moved westward, Walker discovered that by 1890 "the line of the frontier," that is, "a line of continuous settlement," had reached the Pacific Coast. With a burgeoning population the United States had occupied the continent from east to west. There was no longer a frontier line that could "be traced on a map."[48]

What is particularly significant about Walker, aside from his genuine popularity and influence in his own time, is the fact that he was, in effect, the creator of a frontier theory of history which, slightly modified, was later developed by Frederick Jackson Turner. Even Walker's geographical classification of land divisions, or sections, was almost exactly the same as the pattern developed by Turner when he later modified his frontier theory with his sectional hypothesis. Turner's writings and citations to Walker and his heavily lined personal copy of an address by Walker, containing seminal ideas on the frontier theory, are evidence of Walker's compelling influence in helping Turner create a theoretical scaffolding of history.[49] Thus Walker as well as Turner stressed a theory of frontier progress and growth, which in time came to mean the commercialization and conquest of nature. The land, according to this view of historical events, is generally regarded as a commodity instead of a resource because of the stress on conquistador themes of frontier expansionism.[50]

Although in his later years Turner had serious reservations about unlimited population growth and a never-ending exploitation of nonrenewable resources, this line of thought did not reach his published writings.[51] Far ahead of Turner in understanding environmental social costs was a contemporary of his, Thorstein Veblen, son of a Norwegian pioneer who had spent his early years on a farm near Portage, Wisconsin (Turner's home town), where he labored in the fields doing a man's work while still a youth.[52] Considered by some as the greatest frontier voice of his time, Veblen's stream of books and essays dissected the Spencerian theme of Anglo-Saxon elitism in America as well as the historical process of frontier development.

In analyzing America's nineteenth-century dilemma, Veblen concluded that vested interests did not bear their share of environmental costs because the "doing business" rationale of wealthy Americans caused rapid social losses for the nation at large. As an eyewitness to wasteful farming practices and to business domination of government, a situation that permitted the slaughter of buffalo and exploitation of the Indian in his time, Veblen provided a unique and penetrating assessment of what was going on in the United States. The various frontiers of "progress"—the fur trading, mining, ranching, farming, and oil drilling frontiers—Veblen understood as having produced huge social losses, almost impossible to calculate on a monetary basis.[53] Veblen wrote, "This American plan or policy is very simply a settled practice of converting all public wealth to private gain on a plan of legalized seizure." The scheme of converting public wealth to private gain gave impetus, Veblen argued, to the growth of slavery because of the development of one-crop agriculture on a large scale fueled by forced labor. Both agricultural and real-estate speculation were aspects of this progressive confiscation of natural resources. The history of frontier expansion, Veblen maintained, was marked by the seizure of specific natural resources for privileged interests. There was a kind of order to the taking: what was most easily available for quick riches went first. After the despoliation of wildlife for fur trade wealth came, the taking of gold and other precious minerals followed by the confiscation of timber, iron, other metals, oil, natural gas, water power, irrigation rights, and transportation right-of-ways.[54] What was the result of such a shortsighted policy? The inevitable consequence, Veblen maintained, was the looting of the nation's nonrenewable resources to enrich the privileged few. The fur trade, Veblen said, represented this kind of exploitation and was "an unwritten chapter on the debauchery and manslaughter entailed upon the Indian population of the country." The sheer nastiness of this rotten business was such that it produced, according to Veblen, "the sclerosis of the American soul."[55]

Another perceptive witness of the ecological impact of the frontier in the late nineteenth century was John Muir, son of a Scottish pioneer and a person who, like Veblen, labored on a farm as a boy in the Portage, Wisconsin, area. Muir, a critic of the farming frontier of his youth, was an eyewitness to soil despoliation and reckless timber cutting by settlers and land speculators. He also recounted in detail the strange headhunts carried on by whole farming populations at certain times during the year when young people competed in gory contests to fill bloody bags with as many heads of small animals and birds as came within their grasp. All this mayhem, Muir recalled, was in the mistaken belief that exterminating wildlife somehow benefited the farmer.[56]

Muir, it will be remembered, went on to study at the University of Wisconsin and then to travel widely, for as his correspondence shows, he had a burning desire to study, firsthand, the forests and wildlands of the world. His travels in Asia, Alaska, Africa, Australia, South America, and most parts of North America gave him an environmental perspective that was worldwide and made him the most distinguished self-taught naturalist of his day.[57] As an eyewitness, even to the extent of following sheep migrations through the Sierras, he saw more clearly than any other American of his time what his country had lost and what it needed to protect and preserve. As a persistent traveler, with an ever-present notebook that recorded his careful observations, he came to know and understand the dimensions of the cumulative impact of America's westward moving frontiers. There was no doubt in his mind that some of the most magnificent scenic wilderness areas in the world were threatened by herds of sheep and cattle, and by agriculture, mining, timber cutting, dams, and other works of man. In 1889, the year when Veblen published his manifesto on wasteful consumption in America, *The Theory of the Leisure Class*, and Turner was completing his doctoral dissertation that eulogized the fur trading post as a social institution, Muir took the editor of the *Century Magazine* into the Yosemite Valley to show him the damage wrought by invading herds of sheep, the "hoofed locusts."[58]

Muir, as ecologist and founder of the Sierra Club, preached with the oratory of a Hebrew prophet that it was the sacred responsibility of all Americans to prevent their land from being ravaged by the forces of greed and stupidity. Almost single-handedly he sparked the movement to preserve Yosemite as a national park, and partly as a result of his efforts, Theodore Roosevelt set aside 148 million acres as forest reserve. Summing up the environmental impact of the American frontier on the land by 1901, Muir concluded, "none of nature's landscapes are ugly so long as they are wild, but," he added, ". . . the continent's . . . beauty is fast passing away, especially the plant part of it, the destructible and most charming of all."[59]

There are, of course, more recent witnesses of the great despoliation that refined the modern impact of society upon the land and supplement the observations of Muir, Veblen, and Marsh. For instance, in the period of the 1950s Bernard DeVoto, historian and journalist who traveled widely in the West, pinpointed what Veblen had earlier charged: private interests were the most culpable. Especially during the Dwight D. Eisenhower administration, charged DeVoto, they were exploiting the national forests' nonrenewable resources and wildlife. In probing the environmental history of the United States in a series of *Harper's* Easy Chair articles, DeVoto showed how the West had become a "plundered province" where absentee owners acted "on the simple principle: get the money out. And theirs was an economy of liquidation."[60] In an unfinished book, now among his papers at Stanford University, DeVoto described how the American West was, in large part, an arid region that had become an environmental disaster because of stupid land law legislation that ignored the good advice of John Wesley Powell. The inevitable consequence was, DeVoto wrote, a cycle of depressions, bankruptcies, and alienation of the land that caused America's dust bowl of the 1930s.[61] And DeVoto knew about arid lands because he had grown up in Ogden, Utah.

About the time that DeVoto began his environmental crusade through articles, speeches, and an unfinished book, two other conscientious objectors to land exploitation appeared upon the scene. Both had training in the sciences, both were skilled writers, and both of these remarkable men, Carl O. Sauer, a geographer, and Aldo Leopold, a forester, were astute observers, among our best historical eyewitnesses, to what was happening to the land in America and in the world at large.

Sauer, a skilled interdisciplinary researcher in the biological and social sciences and in related fields, argued that man's assault on nature in North America not only brought about the extinction of species but also the severe restriction of other species. This was true among animals as well as plants.[62] Sauer, for example, emphasized our extreme dependence upon commercial corn, which was ecologically dangerous because it increasingly reduced the range of organic evolution to only two of the species first domesticated by Indians.[63]

But it was the soil destruction that Sauer called America's "dreadful problem." Loss of topsoil in many areas had reduced the United States to a deprived nation, Sauer argued. We had dissipated our land wealth. To reconstruct California soil profiles, Sauer had to go to Baja California.[64] And in the American Southeast, Sauer found that as a result of longtime erosion and wasteful agricultural practices, a new subsoil was on the surface, a red soil that could only be kept alive through heavy doses of commercial fertilizer. Such destruction, Sauer maintained, was the outgrowth of our frontier optimism that went back to the days of European freebooters. "We have not learned," he wrote, "the difference between yield and loot."[65]

Sauer's interest in environmental distress led him into brief correspondence with Aldo Leopold, forester and former hunter, who was convinced that America had to take a new look at its vanishing wilderness lands. Leopold, a Wisconsonite[66] who had studied Turner's frontier theory alleging that primitive frontier environments helped to form our national character and democratic institutions, turned Turner upside down with this argument: "is it not a bit beside the point for us to be so solicitous about preserving {American} institutions without giving so much as a thought to preserving the environment which produced them?"[67]

In his *Sand County Almanac* of 1949 and in articles, Leopold developed the idea of "an ecological conscience," "a land ethic."[68] Inherent in his concept was the argument that even carnivores in the wilderness had a role in maintaining the environmental balance. "Just as a deer herd lives in mortal fear of wolves," Leopold wrote, "so does a mountain live in mortal fear of its deer. And perhaps with better cause, for while a buck pulled down by wolves can be replaced in two or three years, a range pulled down by too many deer may fail of replacement in as many decades."[69] Modern Americans, Leopold maintained, like the Spanish conquistadors, "were captains of an invasion," but we were not too sure of our own righteousness. Thus Leopold, forester and pioneer in wildlife management, an eyewitness to the despoliation by bulldozers, farmers, industry, government trappers, and timbermen, concluded that there was need of a land ethic, a moral concept, in formulating policies. We must think about the land, he said, in terms of what is "ethically and esthetically right as well as what is economically expedient."[70]

Leopold wrote eloquently on this point as well as on environmental education, explaining patiently the complexities of ecological balance. There were others who followed him. One was Rachel Carson, a biologist, who in the 1950s began a book on *The Control of Nature*.[71] The title, as we know, was changed to *Silent Spring*, a graceful, articulate volume that even makes the subject of "chlorinated hydrocarbons" fascinating to the reader. She wrote eloquently about chemicals and ecology whereas her predecessors had written about chemicals and economics in the pesticide field. Carson showed that the U.S. government (in particular the Department of Agriculture) and scientists using industry research funds (especially those scientists in the universities and in the pesticide industry) had filled the country with dangerous chemical poisons without having the remotest idea of what they were doing to the environmental balance. The despoliation that resulted from foolish, unsuccessful extermination programs directed at fire ants and gypsy moths was great.[72]

One might conclude, after hearing of the destructive impact of people, industry, and government, that there is not much left of America to be saved. As we know, this is certainly not the case, for America still has a reservoir of natural beauty, productive lands, and a wealth of natural resources. Yet there is no question that today the facts of life tell us that we have a national energy crisis and serious pollution problems in many areas of the country. By looking back over the highlights or themes of environmental despoliation in our history, we can, I believe, agree that there are basic environmental reasons that account, at least in part, for the severity of the problems we now face. These reasons are:

(1) The competitive exploitation of furbearing animals and nonrenewable natural resources greatly exceeded the extent to which this natural wealth had to be sacrificed. And further, the assault upon the land was tied closely to the dispossession of the Indians, our first ecologists.[73]

(2) Pioneers from earliest times wasted natural resources, but it has been the ever-increasing volume of despoliation and its cumulative effects that have brought instances of permanent environmental damage.

(3) The American government has had an increasing role in the despoliation because of its links with predatory business interests and scientists, many of them associated with leading universities. In recent times we can see this trend in certain actions of such

government agencies as the Department of Agriculture, the Army Corps of Engineers, and the Atomic Energy Commission.[74]

(4) Population growth, as predicted by Jefferson, Franklin, and others, has culminated in an occupation of the land from sea to sea. This, with the new affluence, has brought a crowding of the land and tremendous strain upon resources and energy reserves.[75] The ambivalent attitude toward growth and conservation continues, but there is evidence that the old ideal of unlimited growth should be questioned.

(5) Data from interdisciplinary research and from competent eyewitnesses show that the familiar Turnerian frontiers of first contact in the wilderness, fur trading, farming, mining, and urban settlement had reverberating environmental impacts and caused destructive social losses.[76] Most of these losses were unpaid at the time and later generations are therefore obliged to accept the costs in lost capital and in environmental despoliation.

We may also conclude that environmental history, like the "new history" of James Harvey Robinson and Charles A. Beard a generation ago, involves historical lessons and questions of morality. One of these questions is, should we have more respect for the land, even a land ethic? In sort, an examination of historical environmental themes reveals a need for more emphasis upon new attitudes of Americans toward the environment. More stress could well be placed upon historical themes based upon a land ethic and respect for nature as opposed to the old conquistador interpretation. We can also improve our sense of identification with future generations and acknowledge the fact that environmental problems in the United States usually have worldwide repercussions. We look to the past, not to discern what must be, but to understand environmental themes that help to explain origins of ecological transformations taking place in our lifetime. We may then catch a glimpse of our newly forming frontiers that forecast the environmental future of America.

NOTES

1. Charles A. Lindbergh's perspective on American environmental change within his lifetime are, in many respects, representative of many other con-

cerned observers in past decades. Lindbergh, his wife tells us in a recent interview, ". . . flew over the country a great deal and was . . . terrified and horrified by the way civilization and big cities had spread over what was country before. . . . He just felt we were losing our country and our land and . . . the beauty of the land." Anne Morrow Lindbergh, "Conversations with Eric Sevaried," Los Angeles TV, Channel 2, May 27, 1977. John Schutz, Alexander DeConde, Roderick Nash, Calvin Martin, Lynn Donovan, and Robert Brunkow have read and improved this paper with their suggestions.

2. See Alfred W. Crosby's excellent, pathbreaking volume, *The Columbian Exchange: Biological and Cultural Consequences of 1492* (Westport, Conn., 1972), especially his chapters on "Conquistador y Pestilencia" and "Old World Plants and Animals in the New World," pp. 35–121, and his bibliography covering a wide range of interdisciplinary data, pp. 222–260. This should be supplemented by the data and bibliographical notes in traditional Spanish sources in the article, "Antilles," *Enciclopedia Universal Ilustrada Europeo-Americana* (Barcelona, 1905), V, 776–778. Henry F. Dobyns's unpublished paper, "Major Dynamics of the Historic Demography of Indo-Americans," presented at the April, 1977, meeting of the Organization of American Historians, Atlanta, Georgia (copy in my possession), discusses the transmissions of pathogens evolved in the Old World as part of the Columbian exchange that caused a biological catastrophe among native people of the New World. This involved, Dobyns argues, the operation of William H. McNeill's "law of biological aggression" in which conquering societies transmitted their "domesticated" diseases to previously independent, isolated peoples, killing off key leaders and warriors in newly contacted societies. See also William H. McNeill, *Plagues and Peoples* (New York, 1976), 55.

3. In his introductory address, "The Agency of Man on the Earth," Carl O. Sauer, opening a European environmental studies congress honoring George Perkins Marsh, suggested several very useful definitions of environmental history. Sauer's address can be founds in William L. Thomas, ed., *Man's Role in Changing the Face of the Earth* (2 vols., Chicago, 1956), I, 49–69.

4. Columbus's comments (on the first voyage) make his journal more than a seaman's logbook, as Samuel E. Morison has argued in his preface to *Journals and Other Documents on the Life and Voyages of Christopher Columbus,* translated and edited by Samuel E. Morison (New York, 1963), 42. The journal gives us data on flora and fauna, Indians, and geography as well as information on the first contacts. In addition to observing the "very green trees, many streams,

and fruits of different kinds," Columbus at once thought of enslaving these Indians, who, he wrote, "ought to be good servants." See *ibid.*, 64–65, for his comments of October 12, 1492. In his remarks of December 21, 1492, he is more explicit, stating that the Indians would "obey without opposition." *Ibid.*, 128–130.

5. The best account of the destruction of the Bahamas and other Caribbean islands is Bartolomé de las Casas, *Brevíssima relacíon de la destrucción de las Indias* (Sevilla, 1552). This work translated by J[ohn] P[hillips] was printed in England under the title, *Tears of the Indians, Being An Historical and True Account of the Cruel Massacres and Slaughters of above Twenty Millions of Innocent People: Committed by the Spaniards in the Islands of Hispaniola, Cuba, Jamaica, &C., as also in the Continent of Mexico, Peru, & Other Places of the West-Indies, to the Total Destruction of Those Countries. Written in Spanish by Casaus, an Eye-Witness of Those Things—And Made English by J.P.* (London, 1656). This work has been made available in a facsimile printing from an original in the Huntington Library by Academic Reprints, P.O. Box 3003, Stanford, Calif. For specific references to the environmental impact of the Spaniards on the islands, see *Tears of the Indians,* 1–26.

6. *Tears of the Indians,* 3. For a critical discussion of the impact of Las Casas's writings in helping to create a Black Legend of Spanish cruelty, see Philip Wayne Powell, *Tree of Hate: Propaganda and Prejudices Affecting United States Relations with the Hispanic World* (New York, 1971), 30–36.

7. The reliability of Las Casas's estimates of Indian depopulation is also strengthened by the writings of Gonzalo Fernández Oviedo y Valdéz in his *Historia General y Natural de las Indias* (Madrid, Ediciones Atlas, 1959), I, 66–67, where Oviedo, one of the earliest historians of the Americas, states that of the million Indians on Santo Domingo, there "are not now believed to be at the present time in this year of 1548 five hundred persons, children and adults, who are natives and are the progeny or lineage of those first." (Quoted in Crosby, *The Columbian Exchange,* 45.) The key publication in the new Indian demographic studies is Henry F. Dobyns, "Estimating Aboriginal American Population: An Appriasal with a Hemispheric Estimate," *Current Anthropology*, VII (1966), 395–414. For further analysis of the new population estimates, see Wilbur R. Jacobs, "The Tip of the Iceberg: Pre-Columbian Indian Demography and Some Implications for Revisionism," *William and Mary Quarterly,* XXXI (1974), 123–132; Francis Jennings, *The Invasion of America: Indians, Colonialism, and the Cant of Conquest* (Chapel Hill, N.C., 1975), 16–31.

8. Las Casas judged the native population of Hispaniola to be three or four million, but Sherburne F. Cook and Woodrow Borah, in their *Essays in Population History: Mexico and the Caribbean* (Berkeley, 1972), I, 407, argue that the island's population was eight million and that the people of Hispaniola had perfected the production of food plants (maize, beans, cassava) to the extent that their fields had a greater yield per hectare than comparable fields harvested in Europe in 1492. This, along with protein from hunting and fishing, was, according to Cook and Borah, more than enough to feed eight million people. Carl Sauer, in *The Early Spanish Main* (Berkeley, 1966), 68, gives evidence to support Cook and Borah by pointing to the value of Yuca, or cassava bread, as a great staple among the Indians of the Caribbean. According to Las Casas, who had been a commercial farmer, "twenty persons working six hours a day for one month will make a planting of such conuscso that will provide bread for three hundred persons for two years." Sauer cites Las Casas, *Apologética Historia . . .* , chap. 20, an unfinished work, probably written in the 1560s. For a complete list of Las Casas's writings, published and unpublished, much of which bears on Indian demography and early environmental history, see *Enciclopedia Universal Illustrada Europeo-Americana,* XXIX, 913.

9. See Dobyns, "Estimating Aboriginal American population," 410–412; Dobyns, *Native American Historical Demography: A Critical Bibliography* (Bloomington, Ind., 1976), 21–34.

10. The most vociferous of these is Angel Rosenblat, whose conservative estimates of aboriginal population rest on the earlier figures of James Mooney and A. L. Kroeber. For a discussion of Rosenblat's arguments, see Jacobs, "The Tip of the Iceberg," 124–127.

11. For a discussion of Indian ecology (and significant ethnohistorical literature on the subject), see Calvin L. Martin's case study of the impact of Europeans on a northeastern American tribe: "The European Impact on the Culture of a Northeastern Algonquian Tribe: An Ecological Interpretation," *William and Mary Quarterly*, XXXI (1974), 3–26.

12. See Conrad E. Heidenreich, "The Geography of Huronia in the First Half of the Seventeenth Century" (Ph.D. dissertation, McMaster University, 1970), 267–273, which contains a discussion of Iroquoian agricultural technology.

13. *The Ecology of North America* (Urbana, Ill., 1963), 1–3, 17–18, 23–24 ff.

14. Vine Deloria, Jr., *We Talk, You Listen: New Tribes, New Turf* (New York, 1970), 189.

15. Fairfield Osborn, *Our Plundered Planet* (Boston, 1948), 64.

16. James C. Malin, *The Grassland of North America: Prolegomena to Its History* (Lawrence, Kan., 1956), 62 ff., 120–155; Malin, *Winter Wheat in the Golden Belt of Kansas: A Study in Adaption to Subhumid Geographical Environment* (Lawrence, Kan., 1944), 34, 80–82, 84; Malin, *Grassland Historical Studies: Natural Resources Utilization in a Background of Science and Technology* (Lawrence, Kan., 1950), 60 ff.

17. Malin's interest in environmental studies can be traced through his early books: *Indian Policy and Westward Expansion* (Lawrence, Kan., 1921), a work that shows an awareness of Indian lifestyles and the impact that white advance had on Indian culture (see, for instance, pp. 74–75 ff.): and *The United States, 1865–1917: An Interpretation* (Lawrence, Kan., 1924), which attacked wealthy, privileged interests (see page 21). See also the personal copy of this volume owned by Frederick J. Turner, Huntington Library Rare Book No. 246464, with Turner's marginalia dealing with geographical, expansionist themes (pp. 19, 21, 61). In Malin's *On the Nature of History: Essays about History and Dissidence* (Lawrence, Kan. [1952]), written in the later part of his life, Malin attacked Frederick Jackson Turner and Walter Prescott Webb for the geographic-space expansionism theories which tended to ignore the subtle relationships between ecology and history (see pp. 107 ff. where Malin cites his own articles on ecology in *Scientific Monthly* of 1950 and 1952 which had been ignored by historians). In one of his most provocative grassland ecology essays, Malin argued that heavy grassland duststorms were commonplace before the coming of the white man. Malin, "The Grassland of North America: Its Occupance and the Challenge of New Appraisals," in Thomas, *Man's Role in Changing the Face of the Earth,* I, 354–356.

18. Robert Beverley, *The History and Present State of Virginia*, edited with an introduction by Louis B. Wright (Chapel Hill, N.C., 1947), 319. See also pp. 44, 49, 318, for above quotations.

19. Washington is cited in A. F. Gustafson, *et al., Conservation in the United States* (Binghamton, N.Y., 1961), 109. See also John C. Fitzpatrick, ed., *Writings of George Washington* (Washington, D.C. 1931–44) for discussions of erosion and ditching, especially, XXX, 238, XXXVI, 240.

20. Patrick Henry is quoted in Gustafson, *Conservation*, 109.

21. See, for example, *ibid.*, 101–195; Eugene P. Odum, *Fundamentals of Ecology* (3rd ed., Philadelphia, 1971), 24 ff; Avery Odelle Craven, *Soil Exhaustion as a Factor in the Agricultural History of Virginia and Maryland, 1606–1860*

(Urbana, 1926), 163 ff. On page 163 are F.J. Turner's annotations (Huntington Library Rare Book 183897). Here Turner has marked Craven's point that one–crop agriculture, which depleted the soil, was "typical" of "all frontiers."

22. Julian Ursyn Niemcewicz, *Under the Vine and Fig Tree: Travels through America in 1797–1799, 1805 with some Further Account of Life in New Jersey,* translated and edited by Metchie J. Budka (Elizabeth, N.J., 1965), 230, cited in Curtis Solberg, "As Others Saw Us: Travelers in America during the Age of the American Revolution" (Ph.D. dissertation, University of California, Santa Barbara, 1963), 158.

23. William Priest, *Travels in the United States of America, Commencing in the Year 1793 and Ending in 1797, with the Author's Journals of His Two Voyages across the Atlantic* (London, 1802), 11, cited in Solberg, "As Others Saw Us," 157.

24. Cadwallader Colden, *The History of the Five Nations of Canada* (2 vols., New York, 1902), I, 69, 72, 166; W.R. Jacobs, *Dispossessing the American Indian* (New York, 1972), 10, 182.

25. These locations are not publicized because of the danger of renewed hunting and trapping. However, beaver families have been successfully relocated from areas in the Far West to rural parts of Arkansas, according to Arkansas state government reports.

26. I have traveled extensively through almost all of rural New England, including the northern woods of Maine. The great pine softwood forests disappeared long ago, sacrificed for housebuilding, shipbuilding, and other uses. Only a small part of the larger hardwood forests remain, and very recently a part of New England's woodlands have fallen to attacks by pests on birch and elms. In colonial times the extraordinary population invasion of settlers into New England resulted in the wars of conquest against the Indians and rapid occupation of inland areas, particularly river valleys. This phenomenon led to dispossession of Indians, destruction of wildlife, and the rapid depletion of the fragile topsoil cover in large parts of New England by the late 1700s. I have observed many stone fences marking out abandoned fields and pastures dating back to the eighteenth century. Although Indian farmers seem to have maintained the ecological balance in New England, the invasion of settlers who carelessly used the land and introduced herding (Indians had no domestic animals requiring pasture) began a sweeping ecological change. See also Jennings, *The Invasion of America,* 19–20, 61–69, 146 ff., 202 ff.; and Calvin L. Martin, "Keepers of the Game: The Ecological Issue of Indian-White Rela-

tions" (Ph.D. dissertation, University of California, Santa Barbara, 1974 [to be published by the University of California Press later this year]), 150–190.

27. A comment often made by the late John D. Hicks in his American history lectures at the University of California, Berkeley.

28. Robert Brunkow, "Environmental and Social Control in Colonial Rhode Island" (Ph.D. dissertation manuscript to be completed in 1978, University of California, Santa Barbara).

29. *Ibid.*

30. Jeremy Belknap, *The History of New Hampshire* (Boston, Mass., 1792), III, 333–334.

31. Thomas Jefferson, *Notes on the State of Virginia*, edited by William Peden (Chapel Hill, N.C., 1955), 53 ff., 82 ff.; Clarence J. Glacken, *Traces on the Rhodian Shore . . .*(Berkeley, Calif., 1967), 681–682; Jacobs, *Dispossessing the American Indian,* 158.

32. Jefferson, *Notes on the State of Virginia,* 83. Jefferson, however, did have reservations about the possible "spirit, warp, and bias" of foreign immigrants. (*Ibid.*) See also quotations from Jefferson in Albert K. Weinberg, *Manifest Destiny* (Chicago, 1963), 120–121.

33. Henry R. Schoolcraft, *A View of the Lead Mines of Missouri* (New York, 1819, reprinted by the Arno Press, New York, 1972, facsimile edition), 64–67, 249–250.

34. David Dale Owen, *Report of a Geological Survey of Wisconsin, Iowa, and Minnesota and Incidentally of a Portion of Nebraska Territory* (Philadelphia, Pa., 1852), 148–150.

35. *Ibid.*

36. Richard A. Bartlett, *Great Surveys of the American West* (Norman, 1962), 141 ff.; William H. Goetzmann, *Exploration and Empire* (New York, 1966), 327 ff., 430 ff., 478 ff. Goetzmann has an excellent descriptive account of the King, Hayden, and Wheeler surveys. I have made a preliminary examination of the huge mass of data left by these explorers and intend to probe further for information about environmental change.

37. Historiographical controversies concerning the "myth" of the "great American desert" are discussed in Ray A. Billington's excellent bibliographical commentary in Billington, *Westward Expansion, A History of the American Frontier* (4th ed., New York, 1974), 743. See also W. Eugene Hollon, *Lost Pathfinder: Zebulon Pike* (Norman, 1949).

38. John Wesley Powell, *Report on the Lands of the Arid Region of the United*

States, with a More Detailed Account of the Lands of Utah, edited by Wallace Stegner (Cambridge, Mass., 1962).

39. *Ibid.*, 32.

40. One of the popularizers of this kind of thinking, Powell complained, was a Professor Cyrus Thomas who served with the Hayden Survey and later with the Bureau of American Ethnology. Thomas went so far as to argue that farms, mines, towns, and roads led to increased moisture in arid regions: "That is, as the population increases the moisture will increase." Quoted by Powell in *ibid.*, 85. Powell's persistent attempts to get his views before competent audiences are revealed in his correspondence, as for instance, in a letter written in the summer of 1878 (undated) to Professor O. C. Marsh of Yale (Marsh Papers, box 26, Sterling Memorial library, Yale University) enclosing copies of his *Report on . . . the Arid Region* and requesting that his correspondent distribute copies to members of the National Academy of Sciences. He hoped that he would be asked to appear before academy members and to deliver a "statement" providing a "clearer understanding" of the "physical character" of the arid country.

41. George Perkins Marsh, *Man and Nature, or Physical Geography as Modified by Human Action*, edited by David Lowenthal (Cambridge, Mass., 1965), xvii.

42. *Ibid.*, 18–20, 33 ff., 259–260, 304–306, 311–325, 413–414, 447–453.

43. *Ibid.*, 661. Among Marsh's many areas of investigation in connection with man's abuse of the land was the origin of national character and whether or not it was "influenced by physical causes." G. P. Marsh to unknown, June 19, 1847, box 26, Marsh Papers.

44. Udall, *The Quiet Crisis* (New York, 1963), 82.

45. For an excellent analysis of Marsh's *Man and Nature,* see David Lowenthal, *George Perkins Marsh, Versatile Vermonter* (New York, 1958), 246–276. For further comment on Marsh's conservationist views and their impact upon the thinking of geographers, see Lowenthal, "George Perkins Marsh and the American Geographical Tradition," *Geographical Review,* XLII (1953), 207–213. For an appraisal of Marsh's role in awakening the public to urban sprawl and its environmental impact, see Lewis Mumford, "The Natural History of Urbanization," in Thomas, *Man's Role in Changing the Face of the Earth,* I, 388–389.

46. Walker's career is traced in an older and almost worshipful but essentially accurate biography by James Phyinney Munroe, *A Life of Francis Amasa Walker* (New York, 1923); many of Walker's ideas on economic growth are set forth in Walker, *Discussions in Economics,* with an introduction by Joseph Dorfman (2 vols., New York, 1971). This work was first published in 1889.

47. "I must deem any man very shallow in his observation of the facts of life . . . ," Walker wrote, "who fails to discern in competition the force to which it is mainly due that mankind have risen from stage to stage, in intellectual, moral, and physical power." Walker, "Mr. Bellamy and the New Nationalist Party," *Atlantic Monthly,* LXV (1890), 248.

48. For the above quotations, see a clipping (heavily annotated by F. J. Turner) of a Phi Beta Kappa address by Walker given on June 19, 1889, Frederick Jackson Turner Papers, File Drawer 15, Huntington Library. See also Turner's reference to Walker's statements about the census of 1890 in W. R. Jacobs, ed., *Frederick Jackson Turner's Legacy* (San Marino, 1965), 84.

49. Jacobs, *Frederick Jackson Turner's Legacy, 19.*

50. This argument is expanded in W. R. Jacobs, "Frontiersmen, Fur Traders, and Other Varmints: An Ecological Appraisal of the Frontier in American History," *American Historical Association Newsletter,* VII (1970), 5–11.

51. Turner's Papers (in the Huntington Library) for the late 1920s and early 1930s show his great concern for the need to conserve nonrenewable resources through such population control measures as family planning. He reluctantly concluded that educated, middle–class, white families (members of his own class) would be the first to limit sizes of their families, but he believed the program should nevertheless be tested to see if wider population limitations could be obtained.

52. Joseph Dorfman, *Thorstein Veblen and His America* (New York, 1940), 5–10.

53. These ideas are set forth in a number of Veblen's books and articles, but especially pinpointed in one of his last publications, *Absentee Ownership and Business Enterprise in Recent Times: The Case of America* (New York, 1923), 168–169. Veblen's concepts on general "land grabbing" are stated at more length in *ibid.*, 165–186, in a section called "The New Gold."

54. *Ibid.*, 168–169.

55. *Ibid.* John Kenneth Galbraith, in analyzing the impact of Thorstein Veblen's writings upon himself and others, concluded that Veblen was "the greatest voice from the frontier world of America" (Galbraith, *The Affluent Society* [Boston, 1969], 52). When Galbraith was president and program chairman of the American Economic Association, he arranged a program which reflected his own social concerns as well as those of younger, revisionist economists. The papers, in one sense, were reminiscent of chapter headings in Veblen's books and dealt with such subjects as "poor and rich," "contradictions of capitalism," the "women's place" (Veblen wrote at length on women's roles), "arts

in the affluent society," and "the economics of full racial equality." See "Papers and Proceedings of the Eighty–fourth Annual Meeting of the American Economic Association," *American Economic Review,* LXII (May 1971), ii–v.

56. John Muir, *The Story of My Boyhood* (Boston, 1913), 168–169.

57. Muir's voluminous unpublished correspondence at the Huntington Library shows him to be one of the most energetic and widely traveled naturalists of his time. Moreover, though self-taught, he easily communicated with professional scholars and scientists of his day in several areas of knowledge (geology, botany, history) where he was extraordinarily competent. This range of professional skills is also illustrated in the remarkable variety of Muir's publications. See "A Bibliography of John Muir," compiled by Jennie Elliot Doran, *Sierra Club Bulletin*, X (1916), 41–59.

58. Muir is quoted in Roderick Nash, *Wilderness and the American Mind* (New Haven, 1969), 130.

59. Muir, *Our National Parks* (1901; Boston, 1907), 7 ff. Muir in the early 1900s was also concerned about the persistent war against wildlife in America by the human population. Writing to Henry Fairfield Osborn on July 16, 1906, he expressed concern about "the extinction" of animals. "I have written on the subject," he said, "but mostly with no effect. The murder business and sport by saint and sinner alike has been pushed ruthlessly, merrily on until at last protective measures are being called for, partly, I suppose, because the pleasure of killing is in danger of being lost from there being little or nothing left to kill, and partly, let us hope, from a dim glimmering recognition of the rights of animals and kinship to ourselves." William Frederick Badé, *The Life and Letters of John Muir* (2 vols., Boston, 1924), II, 350.

60. Bernard DeVoto, *The Easy Chair* (Cambridge, Mass., 1955), 231–255; DeVoto, "The West: A Plundered Province," *Harper's*, CLXIX (Aug. 1934), 355–364; Gene M. Gressley, *The Twentieth-Century American West: A Potpourri* (Columbia, Mo., 1977), 39–40. See also Robert Edson Lee, "The Easy Chair Essays of Bernard DeVoto: A Finding List," *Bulletin of Bibliography*, XXIII (Sept.–Dec. 1960), 64–69, which identifies all DeVoto's Easy Chair essays.

61. DeVoto, "Western Paradox," container 74, DeVoto Papers, Stanford University Library. DeVoto, in his lecture notes on "Public Domain and Conservation" (container 96, DeVoto Papers), also makes the argument that "gigantic waste & destruction of natural resources, esp. timber," were accompanied by "gigantic graft & corruption" in the period 1850–1870. He placed part of the blame on the land-grant railroads, which "whooped up" their land sales by

propaganda. "Fraud, either deliberate or inadvertent, [was] perpetuated upon hundreds of thousands of people who, upon reaching the great plains in the expectation of bettering their estate, went bankrupt instead: (*ibid.*). See also Gilbert B. Workman, "Only Lovers Can Be Sound Critics: Bernard DeVoto and American Conservation" (M.A. thesis, San José State University, 1960), 80 ff.

62. Sauer, "Theme of Plant and Animal Destruction in Economic History," in *Land and Life: A Selection from the Writings of Carl Ortwin Sauer,* edited by John Leighly (Berkeley, 1969), 148–149.

63. *Ibid.* Sauer's long and tenacious research on the history of the domestication of varieties of maize by Indian people is revealed in his correspondence with the botanist, Edgar Anderson. Their letters detail their discoveries of ancient types of maize. See, for instance, Sauer to Anderson, Oct. 10 and 27, 1941, Aug. 31 and Oct. 8, 1942; Anderson to Sauer, Nov. [?], 1941, Sauer Papers, Bancroft Library, University of California, Berkeley.

64. Sauer, *Land and Life*, 148.

65. *Ibid.*, 154. Sauer was also convinced that modern man did not understand his historic role as "the dominant of ecology" by reason of his being "a fire bringer and then a fire maker." Today's "thermal technics," Sauer wrote, have become "a great hazard" to the world. Sauer to Jacobs, Feb. 21, 1972.

66. Leopold spent his early years in Iowa but became a Wisconsonite by adoption when he was in his thirties. He loved the sand and scrub country near Portage, Wisconsin, the boyhood hometown of Frederick Jackson Turner. Leopold's most productive writings were undertaken when he was a forester at the U.S. Forest Service's Laboratory at Madison. See Nash, *Wilderness and the American Mind,* 182–187.

67. Leopold, "Wilderness as a Form of Land Use," *Journal of Land and Public Utility Economics,* I (1925), 401, quoted in Nash, *Wilderness and the American Mind,* 188.

68. Leopold, *A Sand County Almanac with Essays on Conservation from Round River* (New York, 1974), 237–264.

69. *Ibid.*, 140.

70. *Ibid.*, 262.

71. Frank Graham, Jr., *Since Silent Spring* (Boston, 1970), 21.

72. Carson, *Silent Spring* (Cambridge, Mass., 1962), 162 ff.

73. Although a mass of available evidence exists to show that Indian people should be recognized as our first ecologists (as I have stated earlier in this

paper), there are instances of erosion and other types of environmental stress caused by Indian agricultural practices in ancient Mexico. See, for instance, Sherburne F. Cook, *Historical Demography and Ecology of Teotlalpan* (Berkeley, 1949), 41 ff.

74. For an example of the Department of Agriculture's persistent and reckless support of the use of herbicides, see the report by Fred H. Tschirley, department specialist in physiological ecology, "Defoliation in Vietnam," originally published in *Science*, CLXIII (1969), 779–886, and reprinted in Thomas R. Detwyler, ed., *Man's Impact on the Environment* (New York, 1971), 536 ff. Here Tschirley argues that, although large areas of vegetation "were not defoliated, but were killed," "the concept of defoliating in strips or in checkerboard pattern has great merit" *ibid.*, 544).

For an excellent historical overview of the destructive impact on the U.S. Army Corps of Engineers in America, see Elizabeth B. Drew, "Dam Outrage: The Story of the Army Corps of Engineers," in George E. Frankes and Curtis B. Solberg, eds., *Pollution Papers* (New York, 1971), 189–214. For an example of the Dr. Strangelove attitude toward the environment that the Atomic Energy Commission often exhibited in the 1920s and 1960s, see Glen C. Werth, "Plowshare: Scientific Problems," in William R. Judd, ed., *State of Stress in the Earth's Crust: Proceedings of the International Conference, June 13 & 14, 1963—The Amand Corporation* (New York, 1964), 83–97. Werth in this article is an enthusiastic advocate of the use of nuclear AEC "Plowshare" explosives by the government, industrialists, and ordinary businessmen. "We have a problem. Perhaps," Werth states, "it can be solved by use of nuclear explosive. Let's look into the problem and explore the feasibility." He does acknowledge that this kind of everyday use of nuclear explosives may have to await further developments in weapon technology. Atomic Energy Commission leaders of the "Plowshare" project at one time had plans to explode bombs for a new harbor at Perth, Australia, and a new canal through Central America. Such proposals were dropped because of possible dangerous radiation. Confidential interview, July 1977, Dept. of Geophysics, University of California, Berkeley.

75. E. F. Schumacher, *Small Is Beautiful: Economics as if People Mattered* (New York, 1975), discusses pollution problems and growth, questions of size, proper use of the land, and the importance of reexamining "appropriate" or "intermediate" technologies—that is, technologies less offensive to the environment. See esp. pp. 102 ff. See also Barbara Ward, "Small Is Beautiful," *Los Angeles*

Times, Sept. 20, 1977; and S. David Freeman, *et al., A Time to Choose* (Cambridge, Mass., 1974), 19 ff.; Garrison Wilkes, "Plant Germplasm Resources," *Environmental Review*, I, No. 1 (1976), 2–13.

76. Social losses, or costs, as an economic concept are discussed in K. William Kapp, *Social Costs of Business Enterprise* (New York, 1963), 13 ff. See especially Kapp's analysis of the historic social costs of air pollution, water pollution, and loss of nonrenewable resources (*ibid.*, 47 ff.)

PART II

CONTACT AND ITS CONSEQUENCES

CHAPTER THREE

THE FATAL CONFRONTATION

Early Native–White Relations on the Frontiers of Australia, New Guinea, and America—A Comparative Study*

Wilbur Jacobs has been a pioneer in the study of comparative ethnography. His interest in the field developed from his research on Francis Parkman, the nineteenth-century American historian and comparative ethnographer who used his study of the Sioux Indians to provide insights into the sixteenth- and seventeenth-century Iroquois Indians. Emulating the master, Jacobs journeyed overseas to observe the Aboriginal peoples of Australia and New Guinea in the 1960s. He then compared the resistance of California, Australian, and New Guinea indigenous peoples to European and Euroamerican intrusion and settlement. Jacobs's combination of comparative ethnography and contact history helped open a new and fruitful field of study and generate fresh insights into the clash of Indian and European cultures on the frontiers of North America and the South Pacific.

This study was made possible by research grants from the Huntington Library, American Philosophical Society, Committee on the International Exchange of Persons, and Committee on Research at the University of California, Santa Barbara.

Reprinted from Pacific Historical Review *40 (August 1971), 283–309.*

Most investigators who have given serious study to man in prehistory tell us that Stone Age peoples in many parts of the world evolved a culture that gave them a reasonably satisfying life largely governed by the ecology of their habitat. On the Pacific islands, in Australia, and in North America, although patterns of life varied, native cultures seem to have put little strain on the land and biota. The beliefs and institutions of native people encouraged them to live in balance with the natural resources. For example, there are few instances of native peoples killing off animals that were a part of their food supply. The Pacific islanders who lived in delicate balance with plant and animal life evolved a culture governed by specialized tropical and oceanic environments. To a degree, the same generalization can be applied to the Aborigines of Australia and to many North American Indians. In fact, all three peoples, including the ancestors of the native people of Papua-New Guinea, possessed a remarkable knowledge of plants and animals in their habitat and generally lived a life that today's conservationists would praise highly indeed.[1]

When the virgin areas were "discovered" by Occidentals, however, there were fundamental transformations in the ecological balance of the land which were brought about by superior technology. The native people, in most instances, were pushed aside and their lands were utilized for missions, Sunday schools, mines, plantations, farms, and grazing land. Even well meaning missionaries, in seeking to convert the natives into a labor force, often destroyed the lands and culture of the people they sought to protect.

Of the native people who survived the onslaught of the aliens, few retained their lands or even a portion of them. Many natives were simply killed off by bullets and disease, but some survived to become black- or brown-skinned Europeans. Fewer still maintained their ancient ways and somehow clung to their land. In a sense, the real test of survival was possession of the land. Next to outright extermination the best technique for destroying natives was dispossessing them of their land. Land for the Aborigines was all important because it was a spiritual ingredient of their culture; it determined their social groupings and status; and, finally, it was the source of their livelihood. Thus, if there is a probable test for the survival of a native people it is found in the answer to the question: how much of their land have they retained after the alien invasion?[2]

The three widely separated native people compared here, the American Indians, the Australian Aborigines, and the natives of New Guinea, exhibit striking similarities. All of them had lived for centuries isolated from both western and Asian civilizations. Although North American Indians had copper ornaments, all groups lived in what is called the Stone Age in a social world of clans, tribes, and tribelets, many of them part of larger cultural groups loosely related by ancestral and linguistic ties. All used wooden and stone tools and weapons, and all constructed artifacts for a variety of religious, civic, and utilitarian purposes. All lived in a world of sorcery and medicine men. In their societies, the men mainly occupied themselves with hunting, and in making weapons, tools, and shelters, and in some cases gardening, while women were generally concerned with child care, gathering and preparing food, and often in making mats and baskets.

The history of these people is primarily known to us through the records of explorers, missionaries, government agents, and settlers. All express the white man's point of view and are reflected in our histories.[3] Historical materials are, of course, supplemented by modern anthropological studies.[4] All these people had early relations with Europeans, especially English-speaking Europeans. The Indians of North America, it will be recalled, had their frontiers invaded by waves of Spaniards, Dutch, French, Swedes, and English in the seventeenth and eighteenth centuries, and by Russians, Canadians, and Americans in the nineteenth century. Australia was visited by many European explorers and by Captain James Cook,[5] but the real invasion came when an English penal colony was planted there in 1788.[6] The large island of New Guinea was contacted by Portuguese, Spanish, and Dutch explorers in the early 1700s and by Captain James Cook in 1770.[7] Europeans established small settlements on the coasts of New Guinea in the early 1800s.

The native populations of the three areas had strong religious, social, and economic ties with their lands and all resisted European occupation. Generally, territorial rights of specific native groups were respected by other natives in spite of tribal rivalries and conflicts. Identification with specific areas of land was almost always characterized by sacred landmarks and placenames.[8] These native people lived according to what the conservationist Aldo Leopold has called a *land ethic.* Their sacred rituals, mythology, and religious songs were generally tied to the rhythm of the seasons, the growth of plants and animals, and the ancestral deities who created the world. Groups of natives usually claimed descent from a particular totem, perhaps part animal and part human. Totemism, a concept of life unifying all living things in nature with man as a central figure, had no counterpart in European thought and was seldom understood by the whites who first contacted native people.[9] The Aboriginal's respect for the land emerges from his close spiritual ties with nature. Despite the many variations of Indian, Aboriginal, and native New Guinea religious myths and totemism, all appear to have much in common. The lives these native people led, judged by our standards, were brutal and hard. Yet they succeeded in adapting themselves to their environment and lived in relative harmony with their natural surroundings. They even had certain types of population control.[10] Generations lived and died, seemingly without population explosions or widespread starvation.

Much of New Guinea, especially the western part called West Irian (now part of Indonesia), is practically the same as it was thousands of years ago. Village life is almost unchanged. Before the arrival of Europeans these dark-skinned, wooly-haired people (the Papuan and Melanesian natives differ in stature, hair-texture, and features from each other) had established a subsistence village life based upon gardening, hunting, and fishing. Their domestic animals were the dog, the pig, and the chicken, while staple foods were yams, bananas, and the sago palm. Villages were isolated from each other by precipitous mountains, swamplands, or jungles. Feuding tribes, having no common bond of language, fought for grisly trophies of human heads and flesh. Sorcery, exchange of food and gifts, reciprocal "payback" justice, feasts, dances, and condolence ceremonies for the dead are as familiar to the student of the North American Indian as to the student of the Australian Aborigines.[11] The New Guinea natives, like all other indigenous people of tropical countries, suffer from a variety of diseases that limit population growth. Half of the infants born in the Sepik River interior villages where I visited last year die before their first birthday. Malaria, a repulsive infection called "grilli" that flakes the skin, and hookworm are age-old enemies, but the special gifts of transient Europeans and Asians, leprosy, tuberculosis, and venereal disease, have also taken their toll.[12] Voracious clouds of night-flying mosquitoes force individual natives to crawl into huge straw

Sepic River Grass-Covered House at Sago Village

The Sepic River peoples elevate their homes as insulation against dampness, floods, and insects and use the lower area for storage. Wilbur Jacobs lived in the house pictured above, climbing the log-step ladder on the right to the entrance. (Photograph courtesy Wilbur R. Jacobs.)

A Sago Village on a Small Tributary of the Sepic River

A sacred men's house, where women are forbidden to enter, stands in the distance. In the foreground, a Sepic man chops out the interior of a dugout canoe.
(Photograph courtesy Wilbur R. Jacobs.)

stockings to escape being almost eaten alive. Yet everywhere in Papua-New Guinea the natives appear to live today in reasonable prosperity on their own lands. Mountain villagers own coffee plantations that encircle their gardens and straw-roofed matted buildings. The river and jungle villages in rows of homes, many with second story apartments, are owned by the tribes, which have lived there for centuries. The garden-sorcerer, who produces yams as huge as a man, gardens in a plot used by his ancestors for generations.[13]

Why, we may ask, have the natives of New Guinea been allowed to keep their land when the Australian Aborigines and the American Indians were dispossessed? Why is the native population of Paupua-New Guinea nearly two million while the nonindigenous people (Europeans, Chinese, and others) number only twenty-five thousand? A seasoned British explorer of 1890 put the question this way: "Will he [the New Guinea native] too disappear before the impact of the white race, like the Tasmanian and the Australian Aboriginals have done? I think not," the explorer argued, because "the house-building, horticultural Papuan differs as much from the Australian nomad as the Malay or the Samoan differ from the feebler races they dispossessed."[14] Almost immediately the Europeans realized that they could not evict the New Guinea natives. They were firmly entrenched in their villages, fighting with terrible ferocity when attacked or threatened. Because of the complexity and variety of their language patterns and the diversity of local cultures, no friendly tribe or village could easily be used to conquer others or to aid whites in seizing control of the interior. The harsh climate, the virulent jungle diseases, the forbidding swamps, impenetrable jungles, and steep mountain cliffs gave villagers additional protection. Certainly there was no question of native claims to ownership of the land. Villagers clearly defined their systems of land tenure and garden plots (somewhat similar to those of the Iroquois) through a complex mixture of village, family (descent groups), and individual rights that allowed householders the exclusive privilege of cultivating certain allotments.[15]

Although Dutch, Germans, British, and, later, the Australians acquired limited areas of land for trade and plantations near white settlements, there was never a concerted attempt to take over native land rights. The history of Papua-New Guinea reveals that soon after their 1883 occupation of Papua, the southeastern part of the island, the British decided to prevent dispossession of the natives. The special commissioner's report on New Guinea of 1886–1887 states, for example, that "the land question is no doubt the cardinal one upon which almost everything connected with British policy will turn."[16] Within two years the "protectorate" administration, as it called itself, published regulations clearly outlining basic rules governing native-white relations. Whites were not permitted to purchase land or to have any interest in land obtained from native people, who, in turn, were forbidden to sell to whites. Only the imperial government could negotiate purchases. This discreet policy, along with regulations prohibiting the sale of arms and intoxicants to natives or exploitation of the natives as laborers, helped to prevent native explosions of outrage caused by occasional white abuses. An ordinance of 1888, for example, prohibited "removal" of natives from their own district except "for purposes of education or the advancement of religious teaching."[17]

When the British imperial government formally annexed Papua by a proclamation read at Port Moresby in 1888, some decades after the beginning of British occupation of the area, the Germans had already established themselves in the area immediately to the north, and the western part of the island had been annexed by the Dutch as Netherlands New Guinea. The native people on the coast and on neighboring islands had been exposed earlier to Moslem and European slave recruiters, and to a variety of Christian missionaries, Catholic, Anglican, Lutheran, and Methodist. A large number of the islanders had served as laborers in Queensland or on German Samoa plantations and had brought back bitter reports of exploitation. In the year 1883 there were more than thirty ships recruiting native labor on the shores of New Guinea.[18]

However, the islanders fought back. Papuan ferocity in combat was not easily forgotten. Missionaries suffered incredible hardships. A British commentator in 1886 complained that the native response to "the message of Christ" was that of "a savage race of inhuman murderers."[19] In German New Guinea, white control went no farther than the fringes of settlement. Newspaper files at the library of the University of Papua-New Guinea show the remarkable slowness of the penetration of European government into the wilderness even after Australian authority was extended over the eastern part of the island in 1914. As late as 1931 a Port Moresby newspaper complained that "the remote tribes . . . have hardly been under [European] influence at all."[20] Handicapped by the tropical climate, rugged terrain, and the bewildering complexity of native languages, patrol officers found they had another foe, sorcery, which was, as one visitor put it, "the very air of Papua."[21] When native officers took the field, they sometimes hailed sorcerers before white magistrates for such bizarre offenses as "causing rain to fall at an inopportune time." In a land where there are sometimes two hundred inches of rain a year, the rainmaker could be a formidable villain.[22]

The remarkable perseverance until today of sorcery and such customs as "pay-back" justice indicates how slow the white man's civilization has been to penetrate New Guinea. These people have never really been conquered by Europeans. They have never been deprived of their lands by an encompassing frontier of white settlement. There has been native acculturation but without dominance or assimilation. Like many of the American Indians and many Australian Aborigines, some New Guineans have accepted the white man's money, his religion, his items of dress, and other tangibles of civilization, but they have kept their old customs and rituals. Like some of the Indians and Australian Aborigines, they resisted spiritual domination by the white man. The native people in New Guinea, however, seem to have been moderately successful in maintaining their old ways and a larger degree of self-determination because they still control the ownership of the great mass of land. The original British occupation of the area around the large native village at Port Moresby in 1873 did not go far beyond the double rows of native houses and rows of swaying coconut palms where flourished an extensive trade in pottery, garden products, and sago palm flour.[23]

Today, some hundred years later, Port Moresby is much changed, but the center of the city still has its large native market. It is in this city that the native people have a strong voice in self-government through an elected legislative council.[24]

The history of early native-white contacts on the frontiers of Papua-New Guinea, then, is different from that of North America or Australia largely because the American Indians and the Australian Aborigines were driven from their lands, especially in areas of rich soil where climate sustained lush vegetation. When the native peoples were overrun by advancing white frontiers, as in Australia and North America or on Pacific islands such as New Zealand, investigators have discerned three distinct periods in the acculturation of tribes or groups of Aborigines. In the first stage of white contact, friendly relations often developed because of native hospitality and presents offered by the whites. As whites encroached upon native lands, early friendship soured and within decades native resentment grew. Finally, warfare resulted in defeat of the native and bewildered loss of respect for their own culture.[25]

In the second period, an era of depopulation and despondence that sometimes covered several decades or even a half century, the native people often developed a scorn for many of their own customs. This was usually a time when they suffered from the onslaught of smallpox, tuberculosis, venereal disease, and alcoholism. After portions of their lands and sacred places had been occupied by whites, they were often left without leadership. Finally, there was a stage that the noted Australian anthropologist, A. P. Elkin, calls an era of contra-acculturation.[26] In this period the natives tended to revive their culture in modified form with renewed appreciation of their own arts, crafts, and rituals. If native societies reached this third stage, as the Seneca Indians did with the rise of the prophet Handsome Lake, they usually survived as a people with their own cultural heritage. Otherwise they perished or were assimilated by the whites.

The Seneca took nearly two hundred years to complete this cycle of acculturation. Like the Maoris of New Zealand, who experienced a similar cycle in their confrontations with whites, the Seneca were fierce warriors who had at first held Europeans at bay. The Seneca, in fact, prospered after their initial contacts with the Dutch, French, and British in the seventeenth century. Despite the disappearance of furbearing animals from their country in western New York, they profited from their role as middlemen in the fur trade between the Anglo-Dutch traders at Albany and the Indians of the Great Lakes. They sustained themselves with a diet based upon maize, squash, beans, and nuts, and protected themselves by "the Great Binding Law" or Iroquois "constitution," which bound the Seneca to the Five Nations confederacy.[27]

In their prime during the 1750s, the Seneca, with their fellow-Iroquois, maintained almost a balance of power between France and England.[28] But when the eighteenth-century wars of the forest ended, they lost much of their land. After the American Revolution, they were forced into a reservation life that has been called "slums in the wilderness." The

1799 vision of Handsome Lake, a reformed drunkard and Seneca chief, marked the beginning a new era for his people, a renaissance in Iroquois technology, a rehabilitation of the Iroquois cultural health, and the beginning of the Longhouse or Handsome Lake Church.[29]

Not all Indian prophets had the success of Handsome Lake. The message of a Delaware prophet of the 1750s was taken over by an Ottawa chief, Pontiac, who unsuccessfully led northeastern tribes in one of the last great wars of the forest. These religious movements, which appeared among the plains tribes in the nineteenth century, have sometimes been called "cults of despair." Across the Pacific, full-blood and half-caste Australian Aborigines attempted to revive the rites and beliefs of their ancestors.[30]

Most of the Aboriginals of North America and Australia have worked out a partial assimilation on their own as a result of living within a conquered territory. The Navajo, who stem from the Athapascan Apache, borrowed weaving and planting of corn from the Pueblo tribes, and later silversmithing and sheepherding form the Spaniards. By skillfully adapting these to the arid land left them by the whites, they have increased in population and today are a proud, largely self-sustaining people who have strongly resisted assimilation into white society.[31]

The histories of the Seneca and the Navajo are exceptional in the overall story of Indian-white confrontations on America's frontiers. Many tribes were simply unable to withstand the powerful impact of white frontiers of exploration and settlement. In both Australia and America there was a disintegration of native societies directly adjacent to white settlements, especially in areas where there had been a high native population density.

The highest concentration of native population in both North America and Australia was in areas of moderate rainfall and lush vegetation along coasts and islands of both continents. As A. L. Kroeber repeatedly emphasized, rainfall variations and climatic conditions had a significant influence on the growth of native cultures and population densities. The harvesting of maize by the eastern Indians and the gathering of acorns by the California Indians contributed to the density of Indian population.[32]

Among the coastal Indians of North America and Aborigines of Australia hunting and living areas were usually well defined. These were the areas quickly penetrated by land-hungry whites because of the fertility of the soil and the agreeable climatic conditions that were similar to those in Europe.

The advance of the white frontiers of settlement was generally conditioned by the same factors that originally determined native population patterns and the character of native hunting, harvesting, and food-gathering areas. In great arid regions such as the Aranda Plain of central Australia, a tribal hunting and food gathering territory might embrace 25,000 square miles.[33] Whites generally avoided such areas except for periodic mining booms. Natives of the interior scattered so thinly throughout a large desert zone were not quickly conquered or assimilated. Indeed, a few remain wild and free today. At the time of discovery, there were some three hundred thousand Aborigines in Australia, about one individual for every ten square miles.[34] Before the coming of the whites there were, according to a number of authorities, about one million Indians in what is now the United States, about one for every two to three square miles.[35]

Despite the superior fighting qualities of the Indians and their larger numbers, they were not able to hold off the advancing frontier after the colonial period, and, had they not had the alliance of the French fur-trading interests, it is doubtful if the Anglo-American westward movement would have been held at the Appalachians after the early 1750s.[36] In Australia, a small population of white settlers and convict laborers easily overran in a few decades the best areas occupied by the Aborigines on the southwestern shore of Australia and the large island of Tasmania. New Guinea, by contrast a land of rugged terrain and lush tropical vegetation, supported a native population said to be as high as five hundred per square mile in some areas, a strong barrier to whites who at first barely held their own on the fringes of the island.[37]

The Australian Aborigines who were so easily dispossessed and conquered resembled in many respects the California Indians. These tribesmen, it will be recalled, were easily forced into the Spanish mission system and later left to fend for themselves after the secularization of mission lands and the occupation of California by the advancing American frontier.[38] When A. L. Kroeber in 1954 wrote a paper on "The Nature of Land-Holding Groups in Aboriginal California" in connection with a California Indian land claims case, he made a brilliant summary of tribal use of the land.[39] These Indians were particularly vulnerable to exploitation. Living by hunting, fishing, and acorn and root gathering, they had no art except basketry, no elaborate dances, no cultivated gardens or fields of maize. They lived in rude huts in small communities, and, though often naked, sometimes had rush aprons or skin robes. To the Europeans, their open brush lands might appear useless, but they yielded rabbits, birds, and other animals for food. Forest stands sheltered browsing deer. Each tract, well defined in area, was a source of food to be harvested. Many tribelets moved from area to area in season, fishing in winter and spring, hunting during summer and fall. Like the Aborigines of Australia, they did not tame and breed animals except the dog. Acorns were a primary food, and there was a variety of other wild foods: fruits, berries, seeds, and fleshy root stalks.[40]

These tribelets resembled the subtribes of Australian Aborigines, small food-gathering groups called *hordes*. Unlike the Indians, the Aborigines had no villages; they were seminomadic hunters, collectors, and fishermen who moved across the land. The men stalked the kangaroo or emu or speared fish. The women collected wild fruits, yams, nuts, seeds, grubs, stems, and roots.[41] Each Aboriginal horde's hunting and living territory was clearly defined and respected by other groups. Yet, as one Australian pioneer opined, such people were "strolling savages" without a particular home or habitation.[42]

Other Australian pioneers questioned this approach which tended to justify the dispossession of nomadic people. In 1839 a sympathetic Australian defended the natives and their right to the land:

> In short every tribe has its own district & boundaries [and these] are well known to the natives generally; and within that district all of the wild animals are considered as the property of the tribe ranging on its whole extent [, just] as the flocks of sheep and herds of cattle that have been introduced by adventurous Europeans . . . are held by European law & usage the property of the respective owners. In fact as the country is occupied chiefly for pastoral purposes the differences between the Aboriginal and European ideas of property and the soil is more imaginary than real.[43]

The author of this interesting comparison between Aborigines and Australian settlers, who was probably a Presbyterian clergyman and amateur anthropologist,[44] went on to compare the kangaroos and the wild cattle. "The only difference being," he stated, "that the former are not branded with a particular mark like the latter, & are somewhat wilder and more difficult to catch." Furthermore, he argued, "particular districts are not merely the property of particular tribes, particular sections of these districts are universally recognized by the natives as the property of individual members of the tribes." Finally, he pointed out a factor of great importance: "the infinity of the natives' names of places, all of which are descriptive & appropriate[,] is of itself prima facie evidence of their having strong ideas of property and the soil."[45]

There were few other whites who wanted to consider property rights of the Aborigines in pioneer Australia. Most clergymen, and even the various "protectors" of the Aborigines, usually former preachers who were appointed by the government to round up native tribes and to move them to a new area, had a condescending, fatalistic attitude toward the them.[46] To justify of the occupation of Tasmanian lands, a prelude to complete extermination of the natives (after Lt. Governor George Arthur organized a manhunt across the island in 1830), one clergyman wrote, "It cannot be supposed that providence would desire any country to the occupancy of a few savages who make no further use of it than wandering from place to place when at the same time millions of human beings in other places are crowded upon one another without means of subsisting."[47]

Here, then, we have the two contrasting arguments on dispossession of native people, both made by English clergymen in Australia in the 1830s. Similar arguments were made by those who wished to dispossess the American Indians, as Wilcomb Washburn has pointed out in a learned article.[48] When Theodore Roosevelt wrote that "justice" was on the side of the pioneers because "this great continent could not have been kept as nothing but a game reserve for squalid savages,"[49] he echoed the opinions of such eminent figures in American history as John Winthrop, John Adams, Lewis Cass, John C. Calhoun, and Thomas Hart Benton who argued that a nomadic, primitive race must give way to a Christian, agricultural, civilized society.[50] Much of the justification rested upon Biblical quotations purporting to show that white people had prior rights to the land because they "used it according to the intentions of the creator."[51] The argument that nomadic hunters could be forced to alter their economy by an agricultural or pastoral people had first, though not systematically, been advanced by John Locke.[52]

The nomadic Aborigines of Australia were also held in low esteem by almost all whites from the time of early exploration and settlement. Most

Europeans shared the opinion of the seventeenth-century explorer William Dampier who described their "great bottle noses, pretty full lips, and wide mouths." "Their Eye-Lids," he said, "are always half closed, to keep the Flies out of their Eyes." These people, he said, "differ but little from Brutes," having no houses, domestic animals, fruits, or "skin garments."[53] Anthony Trollope's harsh judgment two centuries later is representative of the prevalent white attitude toward the natives: "Of the Australian black man we may certainly say that he has to go. That he should perish without unnecessary suffering should be the aim of all who are concerned in the matter." Trollope went so far as to write that should the "race" increase it "would be a curse rather than a blessing."[54] He reached this decision despite the remarkable "proficiency" of the children in school. The blacks, he tells us confidently, "are being exterminated by the footsteps of the advancing race."[55] The eminent Anglican clergyman of Australia in the early 1800s, the Rev. Samuel Marsden, held the blacks in such contempt that he felt they were not worth the missionary effort required for conversion. Indeed, their misery, he held, resulted from a special punishment inflicted upon them because of the sins of their ancestors in the Garden of Eden.[56] Marsden, though enthusiastic about missionary work among the Maories in New Zealand, concluded that the Australian blacks "have no Reflections—they have no attachments, and they have no wants."[57]

The comments of Marsden, Trollope, and Dampier on Aborigines span two centuries but illustrate the unchanging contempt for these native people. In the 1790s and early 1800s the Aborigines were quickly killed off in Tasmania and pushed inland from the southeastern shores of Australia. Although log forts were at first built, the farming and pastoral frontiers soon occupied fertile areas of the interior. Great pastoral estates, the squatters' "stations" as they were called, spread into the hinterlands, widely separated from each other because there was no real military threat from the natives.[58] In 1838 Governor Sir George Gipps divided Port Philip in southern Victoria into four districts. He provided a "protector" for the Aborigines after a House of Commons committee reported on the pitiful condition of the vanishing people, degraded by liquor, disease, and poverty. There is evidence that squatters put arsenic into the natives' flour after the "black savages," as they were called, lighted dry grass with fire sticks, ate sheep or cattle, or occasionally attacked and killed isolated whites.[59] An attempt by the British government to halt the extermination of natives by hanging seven whites who had murdered blacks at a settlement near Myall Creek only made whites more secretive about their punitive raids on the natives.[60] The last Tasmanian woman died in 1888, completing the extinction of her race. As late as the 1890s native police in Queensland were engaging in punitive expeditions against other Aborigines.[61] Reservations and missions eventually evolved in the late nineteenth century but failed to stop the decline of the Aboriginal population. As one concerned modern Australian describes the story:

> When our British forefathers took this land they termed it "waste and unoccupied": in reality they conquered the Aboriginal people by force of arms, disease, starvation and the destruction of Aboriginal social systems. We are heirs to a colonial empire which was built largely on force and a deep abiding belief in the superiority of British people and their institutions.[62]

The author of this encapsulated history of native-white relations had studied the same problem in several areas of the world. He concludes that the Australian policy which first permitted occasional extermination, and then, in the twentieth century, encouraged assimilation is based on "racism," "the conscious or unconscious belief in the basic superiority of individuals of European ancestry, which entitles white peoples to a position of dominance and privilege." He reasons that this attitude not only permitted, but encouraged exploitation of native peoples. The policy of assimilation, he argues further, has been based upon ignorance and disdain for the life-style of the Aborigines and is concerned only with turning them into "dark-skinned Europeans."[63]

Similar indictments of European relations with Indians are in William Christy Macleod's fine volume, *The American Indian Frontier*, and in William Brandon's excellent *The American Heritage Book of Indians*.[64] In America the wars of the forest culminated in Pontiac's uprising in 1763, and, after the Indian conflicts of the American Revolution, there was never again any real threat to white expansion into the interior. The idea of a boundary line separating whites from Indians was rapidly accepted and finally became national policy after Calhoun proposed moving the Indians west of the Mississippi into a permanent reserve.[65] That land was penetrated by the massive westward trek of Americans during the 1840s. After fur traders, miners, military leaders, farmers, and railroad builders persuaded the government that Indians belonged on reservations, the land was finally subdivided.

As in Australian history, the native was often portrayed as a nasty nomadic heathen whose beliefs and customs left him virtually a beast. Such anthropologists as the Australian, A. P. Elkin, and the American, Nancy O. Lurie, have pointed out that the Europeans who dispossessed the natives had no appreciation or understanding of native culture, occupational specialization, social control, or economic concepts. White ideas about native illiteracy, sexual mores, modesty, Christian beliefs, and white pride in technological superiority (especially the use of guns) buttressed arrogant European assumptions of superiority.[66] There was no real understanding of the native preference for their own culture and way of life over the European system.

For example, the Reverend John Clayton of Virginia in 1787 described the occupation of the male Indians as "exercise," when he meant hunting. Woman's work, the Reverend Clayton briefly noted, was gardening, mat weaving, pottery making, and cooking.[67] Clayton's comments typify those of hundreds of untrained observers who have given posterity an Indian stereotype: hardworking, industrious women and lazy, pleasure-loving men. This image also emerges from much of the Australian and New Guinea literature on native-white contacts. Indeed, it is the impression this writer had when first visiting Australian Aboriginal reserves (in Central Australia, in Queensland, and in the Darwin area) and isolated villages along the tributaries of the Sepik River in northern New Guinea. My initial observations led me to think that men in Swago village in New Guinea spent almost all their time talking and smoking in their sacred clubhouses, tall, well-proportioned buildings constructed without nails, called the *haus tamberan*. Meanwhile the women, it appeared, worked almost as slaves, caring for children, grinding the pulp of sago palm and straining it to obtain the white cheeselike starchy food that comprises their main diet. But I was mistaken, and later found out that the main occupations of the men in the village were varied, physically exhausting, and complex, resembling those of the male in an American Indian village. The male in the New Guinea village busied himself with hunting, house and canoe building, the making of artifacts, and participation in a variety of colorful civic and religious ceremonies, many of which concerned the governance and cultural life of the village at large.[68]

So it is today among nomadic Aborigines of Central Australia. Similar cultural patterns are found among the Indians of colonial Virginia. According to the anthropologist, Nancy O. Lurie, there were marked similarities between the Indians and the early colonists in male and female division of labor for building houses, hunting, housekeeping, child care, garment making, and cooking.[69] The similarities of both cultures to each other did not lead the white to adopt conciliatory attitudes in the inevitable disputes that arose over occupation of the land. Actually, the stereotype of the Indian in early Virginia history was expanded. Not only was the Indian warrior portrayed as a lazy, pleasure-loving rascal, he was also represented as a treacherous, unclean, pagan savage who ate nasty food and might turn upon one at any time despite his sly professions of friendship.[70] One finds similar stereotypes of New Guinea and Australian Aborigines. In American history there are occasional portrayals of Indian nobility and the natural advantages of native life in the forest by such writers as Roger Williams, and such soldiers and Indian superintendents as Robert Rogers and Sir William Johnson.[71] And in Australian history one finds that early European artists drew pictures of manly, athletic, handsome natives, akin to the noble warrior of early Iroquois and Cherokee portraits. However, later pictures of Australian Aborigines are almost caricatures, depicting little black people with huge heads, ugly faces, and sticklike legs.[72] Both in Australia and in America the contemptuous stereotype seems to have submerged the favorable portrait in order to help rationalize unjust policies.

Misconceptions about both American Indians and Australian Aborigines were based upon ignorance of native culture and its development after it confronted white society. There was little in the European's technology that the Indians in Virginia could not evaluate in terms of their own experience. They had made copper ornaments. As soon as they became accustomed to the noisy blast of firing, they familiarized themselves with guns as well as other metal tools. Similarly, fabrics were a part of native technology, and Indians had already made their own nets, weirs, and garden tools. English ships, to Indians, were much like big canoes. Indian religion was polytheistic, based upon a pantheon, and thus a Christian deity could be added, just as the Indians could adopt the use of guns, needles, and scissors. What whites had difficulty in comprehending was that many of the seventeenth and eighteenth-century Indians viewed themselves as equal to the Europeans. Although they could borrow a deity or a technological innovation, they were for the most part unimpressed with the trappings of civilization.[73] Even those colonial sachems and chiefs who had been to England failed to urge the white man's ways on their fellow tribesmen.[74]

Indians of the colonial period did not consider assimilation a solution to the problem of dealing with the whites. This path would have meant servitude, perhaps slavery, educational programs, adoption of their children, and possibly intermarriage, all proposed by Europeans at one time or another during the colonial era. The Indian solution was to attempt to remove the source of anxiety by direct attack on the whites. Pontiac's aim, in the great Indian war of 1763, was to drive the British frontier back into the sea.[75] The Indians who fought the English during the 1760s were fighting for self-determination. For generations they had governed themselves in their towns and villages. Their fortifications and buildings were impressive, even to the colonists. Gardens, orchards, and grain fields attested to their agricultural skill. Indians had no desire to abandon their own culture. The Cherokee who made a last stand against the English in 1760 were such expert farmers that William Wirt, their legal defender in the later controversy over their removal to the far West in 1830, found whites fearful that Indian skills would prevent Georgia's occupation of Cherokee land. The Georgians insisted upon regarding the Cherokee as hunters and argued that these Indians "had no right to alter their conditions to become husbandmen."[76] This remarkable argument denies, of course, the right of the Cherokee to become farmers within the area of Georgia territory. Georgia's argument was carried forward until "it was made clear that, though the Georgian soil was destined to be tilled, it was destined to be tilled by the white man and not the Indian."[77] Thus, white racist rationale was employed to justify taking lands occupied by the Cherokee, one of the great Indian peoples who had tilled the land for centuries. Dispossession of nomadic plains Indians or seizure of lands occupied by the more populous California Indians required no such Byzantine reasoning. Whites could call upon John Locke and the "Creator" to justify their land grabs.

At least the early record of Canada's in dealings with the Indians is notably better than that of the United States. The fur trading tradition of the Hudson's Bay Company and Montreal required that understanding and friendship be maintained with the Indians if the resources of the land were to be successfully exploited. Thus in Canadian relations with the Indians, private or exclusive occupation of the land was not necessary or even desirable in most areas.[78] In both the United States and Australia, however, the persistent clamor of the pioneers for land was the basic factor in the desire of frontier settlers to rid themselves of aboriginal people.[79]

In New Guinea, as we have seen, the indigenous people were never conquered and dispossessed of their lands. They were saved from the worst evils of white racism and possible decimation. Geography, climate, and the physical and cultural vitality of these and other native people all played a part in determining the fate of individual tribes in their confrontation with whites. Europeans, sometimes brutalized by their own cruelty toward Aborigines when the opportunity for gain was present, have yet to seek accommodation. This may be found in understanding the land ethic of the native people, which allowed them to depend upon an economy closely governed by the ecology of the surrounding wilderness.

Yet nations today that are led by a powerful white citizenry are still often unsympathetic to the aspirations of native peoples. For instance, the popularity of Frederick Jackson Turner's frontier theory in North America and in Australia is evidence of the historians' concern for the development of white civilization and the exploitation of the land. Native peoples play only a minor role in this widely accepted interpretation. Turner, in his influential essay of 1893, dismissed the Indian as "a consolidating agent" who helped to encourage intercolonial cooperation for border defense.[80] Turner also treated Indians in his lectures as if they were some kind of geographical obstacle to the westward movements of whites.[81] Australia and British writers who have applied the Turner theory to Australian history sometimes equate the Blue Mountains with the Blue Ridge range and the bushranger with the mountain man, but they are hard pressed to explain the Australian character as an outgrowth of occasional conflicts with the Aborigines, because the fighting was so completely one-sided.[82] A leading Australian historian, tongue-in-cheek, tells us that the timid, peaceful Aborigines may have helped to bring a friendly, law-abiding society to Australia, where whites usually settle their quarrels without violence.[83] Be that as it may, Indians and Aborigines have surely influenced the course of history in America and Australia more than is often recognized, if only by giving a special tincture to the society of the whites who occupied their lands. There are no writers at the University of Papua-New Guinea who see Turner's frontier theory as applicable to that country, for it largely belongs to the indigenous people. The frontier theory, an interpretation of the development of white characteristics in a new land, cannot be applied to a country where the natives still control the mass of their own land and outnumber the Europeans.

The frontier theory, then, represents not only an interpretation of history but also an attitude that historians have taken toward the land, native people, and the expansion of white civilization. As has been mentioned, the white attitude toward native people is often a powerful factor in determining governmental policies. In Papua-New Guinea today, the Australian government, under the spotlight of the United Nations and as trustee for an underdeveloped country, has made great strides in giving the native people what might be called a "fair deal" in social services, land policies, and in self-government. This accords with the attitudes of many enlightened Australians whose influence has been so powerful that Australian public opinion now supports fair treatment for the Aborigines at home.[84]

The fact that the New Guineans still retain most of their land and cultural heritage is probably due to chance, a fortunate combination of circumstances that enabled them to withstand the frontier of white advance. A different set of circumstances allowed European greed to prevail over tolerance on the frontiers of Australia and North America. If we condemn the white man in his relations with the Indians and the Australian aborigines, we must be aware that the same kind of abominations can occur again if the stakes are high enough. A case in point is the future policy of the United States in recognizing oil, mineral, and land claims of Alaskan Indians and Eskimos. Americans have not created a society in which greed is not a controlling motivation. If different races are to live together in harmony in a pluralistic society, then they must free themselves from the urge to look down upon what may seem to be an inferior way of life.

We must also try to understand nativist movements that are oftentimes supported by people whose older values have been lost and whose new ambitions are difficult to achieve. In Australia, in New Guinea, and among the American Indians there are such activist native movements. We should be reluctant to portray these activities as Communist inspired for there is a fundamental difference between the ancient native communalism and modern Marxist Communism. We must realize that modern nativist activism has its basis in a real disagreement with a white man's culture that has taken so much and given so little.[85]

NOTES

1. Douglas L. Oliver in his *Pacific Islands* (3rd ed., New York, 1961), ix, 1–80, stresses the importance of ecological factors in shaping the cultures of native peoples of the Pacific islands, New Guinea, and Australia. Similarly, Alfred L. Kroeber, Frank G. Speck, John M. Cooper, and William F. Fenton have described Indian cultures that maintained an ecological balance with wilderness areas. Speck, for example, in *The Penobscot Man* (Philadelphia, 1940), 207 ff., analyzes the family hunting-ground system of the northeastern Indians, a method of conserving beaver supply. Debate on Speck's theories and discussion of the impact of the fur trade on the northeastern Indians is in Rolf Knight, "A Re-examination of Hunting, Trapping, and Territoriality among the Northeastern Indians," in Anthony Leeds and Andrew P. Vayda, eds., *Man, Culture, and Animals* (Washington, D.C., 1965), 27–41. Calvin Martin, in an unpublished paper, "The Algonquin Family Hunting Territory Revisited" (University of California, Santa Barbara) shows that evidence in the *Jesuit Relations* supports Speck's theories on Indian conservationist techniques. Further discussion on this point is in Eleanor Leacock, "The Montagnais 'Hunting Territory' and the Fur Trade," *American Anthropological Association Memoir No. 78* (Beloit, Wisconsin, 1954) 24–40. The remarkable agricultural techniques of the Hurons who successfully maintained productive corn fields for a dozen years or more are analyzed in Conrad E. Heidenreich, "The Geography of Huronia in the First Half of the 17th Century" (Ph.D. dissertation, McMaster University, 1970), 267–273. Paleoindians and other Paleo peoples have sometimes been portrayed as destructive. For instance, the controversial hypothesis of the Pleistocene "overkill" of huge Ice Age mammals by certain Paleoindians some twelve thousand years ago is examined from several sides by paleontologists, archaeologists, and ecologists in Paul Martin, ed., *Pleistocene Extinction:*

The Search for a Cause (New Haven, Conn., 1967), 75 ff. The question of "over-kill or overchill" is still unresolved. The controversy concerning the Indian and other native people as practitioners of burning (and thereby destroyers of flora and fauna) is discussed in Carl Sauer, *Land and Life* (Berkeley, 1965), 189–191. Sauer, discussing the matter in a letter to me of January 28, 1971, writes: "I think the case is pretty well made for man, and especially the Indians as practices of burning. So the longer the Aborigines and successors were around the bigger the grasslands and the more open the woodlands, the greater the number and diversity of . . . flowering, palatable plants. On balance this meant that there was more food than in fire free tracts. A forest has little food except at the tree tops and along the openings. Indian burning did change faunal composition but increased productivity of food of plant and animal."

2. See Douglas Oliver's perceptive essay on "The Dispossessed" in *The Pacific Islands,* 157–173.

3. Edward H. Spicer, in *Cycles of Conquest: The Impact of Spain, Mexico, and the United States on the Indians of the Southwest,* 1533–1960 (Tucson, 1962), 581–582, discusses, for example, H. H. Bancroft's idea that "savages cannot be civilized under the tuition of superior races," a point of view found in the sources Bancroft used.

4. The literature published by anthropologists on these native peoples is immense. A selection of the main publications consulted for this paper includes A. L. Kroeber's superb volume, *Anthropology, Race, Language, Culture, Psychology, Prehistory* (New York, 1948), which digests a large mass of data published before 1948. The best work on native Australians is A. P. Elkin's classic, *The Australian Aborigines: How to Understand Them* (3rd ed., Sydney, 1954). This is supplemented by a series of research studies by Elkin's former students and admirers, *Aboriginal Man in Australia, Essays in Honour of Emeritus Professor A. P. Elkin*, edited by Ronald M. Berndt and Catherine H. Berndt (Sydney, 1965), and an excellent paperback by R. M. and C. H. Berndt, *The First Australians* (Sydney, 1969). *The Australian Aborigines*, a short illustrated volume published by the Department of Territories (Sydney, 1967), tends to give a favorable coloration to government policies. D. J. Mulvaney, ed., "Australian Archaeology, A Guide to Field Techniques," *Australian Institute of Aborginal Studies Manual No. 4* (Canberra, 1969), 119–130, has a classification of Aboriginal stone implements. A basic research tool is Felix M. Keesing's comprehensive *Culture Change, An Analysis and Bibliography of Anthropological Sources to 1952* (Stanford, 1953).

One of the best modern studies on the native people of New Guinea is by a political scientist, C. D. Rowley, whose *The New Guinea Villager* (Melbourne, 1965) is based upon firsthand experience. *Pigs for the Ancestors: Ritual in the Ecology of a New Guinea People*, by Roy A. Rappaport (New Haven, 1967), is a technical study showing how close the ritual cycle of native people in Tsembaga governed their adjustment to their environment (see especially pp. 224–242). The Tsembaga people keep four kinds of animals: pigs, chickens, dogs, and dassowaries; the tame birds are captured as chicks and provide meat and feathers. Other useful volumes on the New Guinea villagers are: H. Ian Hogbin, *Transformation Scene: The Changing Culture of a New Guinea Village* (London, 1951); Gavin Souter, *New Guinea: The Last Unknown* (London, 1963); Brian Essai, *Papua and New Guinea: A Contemporary Survey* (London, 1961); and *Studies in New Guinea Land Tenure: Three Papers,* by Ian Hogbin and Peter Lawrence (Sydney, 1967). An excellent survey of the island and its native people by two geographers is *New Guinea: The Territory and Its People*, by D. A. M. Lea and P. G. Irwin (Melbourne, 1967).

Basic for the study of the North American Indians are Frederick Webb Hodge, ed., *Handbook of the American Indian North of Mexico* (2 vols., Washington, D.C. 1907, 1910); Hodge, ed., *Handbook of Indians of Canada* (Ottawa, 1913); A. L. Kroeber, *Handbook of the Indians of California* (Washington, 1925); Kroeber, *Cultural and Natural Areas of Native North America* (Berkeley, 1939); John R. Swanton, *Indian Tribes of North America* (Washington, 1953); and S. F. Cook's statistical studies on *The Conflict between the California Indian and White Civilization* (4 vols., Berkeley, 1943), supplemented by the excellent bibliographical notes in Wendell H. Oswalt, *This Land Was Theirs: A Study of the North American Indian* (New York, 1966). William Brandon's *The American Heritage Book of Indians* (New York, 1961) is a readable history of the Indians, brilliantly incorporating a large mass of anthropological data.

5. J. C. Beaglehole, ed., *The Journals of Captain James Cook on His Voyages of Discovery, the Voyage of the Endeavor* (Cambridge, England, 1955), 508, includes a passage from a letter writeen by Cook in 1771 which gives a kind of noble savage image of mainland Australian Aborigines: "These people may truly be said to be in the pure state of Nature, and may appear to some to be the most wretched upon the Earth: but in reality they are far more happier than . . . we Europeans, [since,] being wholly unacquinted [sic] . . . with the superfluous . . . necessary Conveniencies so much sought after in Europe[,] they are happy in not knowing the use of them. . . ." Cook and other writers stressed that the

Aborigines seemed happy living on only the bare necessities, yet Europeans were unhappy. There was an undercurrent of dissatisfaction with civilization because it had abandoned nature. Alan Moorehead in his perceptive book, *The Fatal Impact, An Account of the Invasion of the South Pacific, 1767–1840* (Ringwood, Victoria, Australia, 1966), 150–151, comments on this attitude of certain Pacific explorers.

6. C. M. H. Clark, ed., *Select Documents in Australian History, 1788–1850* (Sydney, 1969), 43 ff.

7. Donald Craigie Gordon, *The Australian Frontier in New Guinea, 1870–1885* (New York, 1951), 19–42, summarizes activities of first explorers in New Guinea. See also Andrew Sharp, *The Discovery of Australia* (London, 1963), 21 ff.

8. A. P. Elkin's chapter, "The Land and the Aborigines," in his *The Australian Aborigines*, 24–48; essay on Aboriginal land rights dated Nov. 15, 1839, possibly written by John D. Lang (1799–1878), clergyman and early anthropologist, in manuscript volume labeled "Aborigines," A 610, Mitchell Library, Sydney; *Indian Place Names, Their Origin, Evolution, and Meaning* by John Rydjord (Norman, Oklahoma, 1968); Erwin G. Gudde, *California Placenames: The Origin and Etymology of Current Geographical Names*(2nd ed., Berkeley, 1960), and the linguistic and ethnographic criticism of this book by William Bright in the *American Journal of Folklore,* LXXV (1962), 78–82; A. L. Kroeber, *California Placenames of Indian Origin* (Berkeley, 1916), diagram 2; "Religious Beliefs and Change of Land Rights," in Hogbin and Lawrence, *Studies in New Guinea Land Tenure,* 117.

9. Elkin, *The Australian Aborigines*, 132–155; R. M. and C. H. Berndt, *The First Australians*, 74–78; Kroeber, *Anthropology*, 396; William N. Fenton, "The Iroquois in History," paper read at the Wenner–Gren Symposium, Burg Wartenstein, Austria, August 7–14, 1967; Hodge, *Handbook of the American Indian,* part II, 787–795. New Guinea natives' shamanism, magic, and animistic ritual is described in Rappaport, *Pigs for Ancestors*, especially "Pigs, Eels, and Fertility," 210–213.

10. Benjamin Franklin described Iroquois population control in this way: "The number of savages generally does not increase in North America. Those living near the Europeans steadily diminish in numbers and strength. The two sexes are of a cold nature, for the men find that the women refuse to sleep with them as soon as they become pregnant. For they believe that makes childbirth difficult. Further, they suckle their children for two and a half or three full years and for the whole time they refrain from sleeping with men." Leonard

W. Labarec, ed., *The Papers of Benjamin Franklin* (New Haven, 1969), xiii, 351. For a modern appraisal of this method of limiting population growth, see Christopher Tietze, "The Effect of Breastfeeding on the Rate of Conception," *Proceedings of the International Population Conference* (New York, 1961), II, 129–136. Among Dr. Tietze's conclusions are: "Since . . . breastfeeding tends to prolong the interval between pregnancies, it seems worthwhile to evaluate it as a method of child spacing" (p. 133).

11. Rowley, *The New Guinea Villager*, 32–52; P. Biskup, B. Jinks, and H. Nelson, *A Short History of New Guinea* (Sydney, 1968), 1–28. The variety of modern native life is described by a former patrol officer, J. P. Sinclair, in his *Behind the Ranges, Patrolling New Guinea* (Melbourne, 1966).

12. The specific new diseases that caused depopulation among Pacific islanders and Australian Aborigines after contacts with Europeans are discussed in Felix M. Keesing, *The South Seas in the Modern World* (New York, 1941), 57 ff., 367; Kroeber, *Anthropology*, 182 ff. Kroeber makes the point that epidemics, once so deadly to islanders and American Indians, are reduced to the level of mild virulence after a generation or two. Edward Spicer stresses a cycle theory of the conquest and withdrawal of Europeans which may leave behind newly invigorated native societies much enriched by cultural exhange. Spicer, *Cycles of Conquest,* 568.

13. Rowley, *The New Guinea Villager,* 115.

14. Theodore F. Bevan, *Toil, Travel, and Discovery in British Guinea* (London, 1890), 276.

15. Hogbin and Lawrence, *Studies in New Guinea Land Tenure*, xiii, 32–33, 100–134.

16. Statement by John Douglas, "Her Majesty's Special Commissioner to New Guinea," Dec. 31, 1886, *British New Guinea, Report for the Year 1886* (Victoria, 1887), 8. A collection of early British reports on the protectorate is preserved in the New Guinea History Collection, library of the University of Papua–New Guinea, Port Moresby.

17. *British New Guinea Annual Report, Her Majesty's Administrator of Government from 4th of September, 1888, to 30th June 1889* (Melbourne, Victoria, 1890), 6.

18. Biskup *et al., A Short History of New Guinea,* 25. This labor trade, sometimes called "the blackbirding trade," often differed little in practice from slaving. Rowley, *The New Guinea Villager*, 58.

19. *British New Guinea Report for the Year 1886,* pp. 8–9.

20. *Papuan Courier*, June 16, 1931.

21. *Ibid.* Perceptive comments on native sorcery are in Beatrice Grimshaw, *The New Guinea* (Philadelphia, 1911), 200; Hogbin, *Transformation Scene: The Changing Culture of a New Guinea Village*, 136, 142–147, 222–226.

22. Lea and Irwin, *New Guinea: The Territory and Its People,* 18–19, chart rainfall up to 250 inches per year in parts of New Guinea and New Britain.

23. The gradual changes of the "salubrius" native village of 800 native people located at the site of Port Moresby is described in *Australasia,* by Elisée Reclus, edited by A. H. Keane (London [1889]), 313–314.

24. Brian Essai, *Papua and New Guinea: A Contemporary Survey* (Melbourne, 1961), 237–238; Albert Maori Kiki, *Kiki, Ten Thousand Years in a Lifetime, A New Guinea Autobiography* (Melbourne, 1968), 161–187, gives the viewpoint of an elected leader of the New Guinea native people. See also John Wilkes, ed., *New Guinea . . . Future Indefinite?* (Sydney, 1968), 139–167.

25. A. P. Elkin has criticized the missionaries who had good intentions but nevertheless were destructive pioneers of white civilization in breaking down native cultures without making replacement. His strictures are in *The Australian Aborigines,* 29, 44, 156–162 ff., and are echoed in A. Grenfell Price, *White Settlers and Native Peoples . . .* (Melbourne, 1949), 194. Missions appear to have created a cycle of destruction that is typically uniform among both the Australian Aboriginals and the California Indians, causing the virtual disappearance of full-blood native people within fifty years. Price, *White Settlers and Native People,* 194–195, and Cook, *The Conflict Between the California Indian and White Civilization: The Indian Versus the Spanish Mission,* 3–12, 15, 113–128.

Elkin's discussion of "Phases in Aboriginal-European Contact" is in *The Australian Aborigines*, 321–328, and in a slightly revised form in Price, *White Settlers and Native People*, 196. A similar discussion is in Felix M. Keesing, *The South Seas in the Modern World*, (New York, 1941), 79–80. Keesing's study of *The Menomini Indians of Wisconsin: A Study of Three Centuries of Cultural Contact and Change* (Philadelphia, 1939) reveals a recognizable tribal identity among these Indians, but only fragments of their Aboriginal culture survived. The Senecas were relatively successful in preserving elements of their culture according to Anthony F. C. Wallace in his *The Death and Rebirth of the Seneca* (New York, 1969), 303–337.

26. Elkin, "The Reaction of Primitive Races to the White Man's Culture," *Hibbert Journal*, XXXV (1937), 537–545, cited in Price, *White Settlers and Native People*, 196, 225.

27. William N. Fenton, ed., *Parker on the Iroquois, Iroquois Uses of Maize and Other Food Plants, The Code of Handsome Lake, the Seneca Prophet, the Constitution of*

the Five Nations (Syracuse, New York, 1968), Introduction, pp. 25–47; Book One, pp. 5–113; Book Three, pp. 7–132.

28. Wilbur R. Jacobs, *Wilderness Politics and Indian Gifts, The Northern Colonial Frontier* (Lincoln, Neb. , 1966), 5, 159 ff. An excellent documented account of the dispossession of Iroquois tribal communities is in Georgiana C. Nammack, *Fraud, Politics, and the Dispossession of the Indians, The Iroquois Land Frontier in the Colonial Period* (Norman, Oklahoma, 1969), 22–106.

29. Fenton, ed., *Parker on the Iroquois,* Book Two, pages 5–138; Wallace, *The Death and Rebirth of the Seneca,* 239 ff.

30. Elkin, *The Australian Aborigines*, 328–338. For the relationsip between the Delaware Prophet and Pontiac, mentioned above, see Wilbur R. Jacobs, "Was the Pontiac Uprising a Conspiracy?" *Ohio Archaelogical and Historical Society Quarterly,* LIX (1950), 26–37.

31. Kroeber, *Anthropology,* 431. Kroeber makes the interesting point that, if by a miracle a major Indian tribe had conquered the whites, our culture would be perhaps only slightly modified (p. 430).

32. Kroeber, *Cultural and Natural Areas of Native North America,* 46, 52, 206–228.

33. Australian Department of Territories, *The Australian Aborigines* (Sydney, 1967), 6.

34. *Ibid.*, 3.

35. By 1900 the native Indian population had dwindled to some 250,000. Estimates of the modern Indian population range from 450,000 to 550,000. The Indian population is expected to go above 700,000 by 1975 according to William Brandon in his *The American Heritage Book of Indians,* 360. The current debate on Indian population estimates is highlighted by Henry F. Dobyns's statistical study, "Estimating Aboriginal American Population. An Appraisal of Techniques with a New Hemisphere Estimate," *Current Anthropology*, VII (1966), 395–449. Dobyns's startling conclusion is that "the New World was inhabited by approximately 90,000,000 persons immediately prior to discovery." He estimates the Indian population of North America at 9,800,000! (pp. 415–416). If Dobyns's estimates are valid (and they are carefully evaluated by qualified scholars who offer critiques in accompanying pages in the article), Europe (with a population of some 100,000,000) may well be considered an invader of new lands with native peoples who had almost the same population. Dobyns believes that epidemic diseases caused the widespread Indian depopulation that occurred soon after white contacts (pp. 412 ff).

36. Indians lost their importance as a balance of power after 1763. W. R. Jacobs, "British-Colonial Attitudes and Policies Toward the Indian in the American Colonies," in H. H. Peckham and Charles Gipson, eds., *Attitudes of the Colonial Powers Toward the Indian* (Salt Lake City, 1969), 81–106.

37. Biskup *et al., A Short History of New Guinea*, 8, a Chimbu area estimate.

38. The deplorable condition of many of the California Indians after American occupation is described in *The Indians of Southern California in 1852: The B. D. Wilson Report and a Selection of Contemporary Comment,* edited by John W. Caughey (San Marino, Calif., 1952). See also Cook, *The Conflict between the California Indian and White Civilization, The American Invasion, 1848–1870*, pp. 5–95.

39. Kroeber, "The Nature of Land-Holding Groups in Aboriginal California," in *Aboriginal California: Three Studies in Cultural History* (Berkeley, Calif., 1963), 81–120.

40. *Ibid.*; William Brandon, "The California Indian World," *The Indian Historian*, II (Summer, 1969), 4–7.

41. Elkin, *The Australian Aborigines*, 1–23.

42. "I confess myself at a loss to comprehend how a few strolling savages, entirely ignorant . . . [and] averse to cultivating the land[,] . . . may be said to possess a small portion of it today by erecting their crude huts, [since they] will abandon it tomorrow. . . ." Diary of Mary Thomas, p. 185, quoted in Ralph M. Hague, typescript MS, "The Law in South Australia, 1836–67," chap. 11, p. 3, State Library of South Australia, Archives Dept., Accession No. 1051.

43. [John D.] Lang, essay in manuscript volume labeled "Aborigines," A 610, Mitchell Library. A portion of the signature on this document is obscured.

44. *Ibid.* See also note 8.

45. Lang, "Aborigines," A 610, Mitchell Library.

46. See, for example [George Augustus] "Robinson's Reports," on the Tasmanian Aborigines," A 612, pp. 76–77, Mitchell Library, Sydney; see especially the letter dated Sept. 9, 1829, from Robinson to Governor George Arthur, in which Robinson comments on "the mortality" that "has pervaded the whole Aboriginal population." Robinson wrote that after contact with "white men" the Aborigines were "imbibed [with] similar debauched habits and vicious propensities."

On the creation of the "protectorate" system by the British government in 1838–1842, see Box 3, "Aboriginal Protectorate," State Library of Victoria Archives, Melbourne. A "Chief Protector" was appointed with four assistant protectors, each officer to have a "district." He was to induce natives "to assume more settled habits of life . . . and watch over the rights and interests of the natives . . . " These officers had duties somewhat similar to those of the Indian superintendents in the British colonies. See W. R. Jacobs, ed., *The Appalachian Indian Frontier, The Edmond Atkin Report and Plan of 1755* (Lincoln, Neb., 1967), xvi ff.

47. Comment by Thomas Henry Braim (1814–1891), Anglican clergyman of Hobart and Sydney in a handwritten essay, "The Aborigines," pp. 3–4, A 614, Mitchell Library.

48. Washburn, "The Moral and Legal Justifications for Dispossessing the Indians," in James M. Smith, ed., *Seventeenth-Century America: Essays in Colonial History* (Chapel Hill, 1959), 15–32.

49. Quoted in *ibid.*, 23.

50. See Albert K. Weinberg's well-documented chapter, "The Destined Use of the Soil," in *Manifest Destiny, A Study of Nationalist Expansionism in American History* (Baltimore, 1935), 72–99.

51. Statement by Senator Thomas H. Benton. Quoted in *ibid.*, 73.

52. Washburn, "The Moral and Legal Justifications for Dispossessing the Indians," 23.

53. Dampier is quoted at length in C. M. J. Clark, *A History of Australia, From the Earliest Times to the Age of Macquarie* (Melbourne, 1962), 39–40.

54. Anthony Trollope, *Australia*, edited by P. D. Edwards and R. B. Joyce (St. Lucia, Queensland, 1967), 113, 475.

55. *Ibid.*, 475.

56. Clark, *A History of Australia,* 169.

57. Quoted from A. T. Yarwood's penetrating essay on Marsden in *Australian Dictionary of Biography* (Melbourne, 1967), II, 209.

58. Geoffrey Blainey, *The Tyranny of Distance, How Distance Shaped Australia's History* (Melbourne, 1969), 132.

59. Aboriginals were often referred to as "blacks" or "savages" in newspaper accounts of disturbances and occasional murders of settlers. See, for example, the *Melbourne Argus,* Nov. 20, 1846, p. 2, for an account of troopers capturing three natives, Bobby, Tolmey, and Bullet-eye, all charged with murder. "Poisoning" Aborigines "was fairly widespread" according to Bryan W. Harrison, author of a B.A. honors thesis, "The Myall Creek Massacre and its Significance in the Controversy over the Aborigines During Australia's Early Squatting Period" (New England University, Armidale, 1966), 101–102. On page 102 n., Harrison cites the following publications as evidence of poisoning: *The Colonist,* July 4, 1838; *Sydney*

Gazette,, Dec. 20, 1838; *Sydney Monitor*, Dec. 24, 1838. The Myall Creek massacre and white use of poisons are briefly covered in C. M. Clark, *A Short History of Australia* (New York, 1963), 87. See also Kathleen Hassel, *The Relations Between the Settlers and Aborigines in South Australia, 1836–60* (Adelaide, 1966), 2 ff.

Whites were reluctant to acknowledge use of poisons against Indians, but, according to Jim Mike, a Ute Indian interviewed on June 20, 1968, the Ute chief, Posey, "was poisoned with flour" in 1923. Doris Duke Tape, 550, Western History Center, University of Utah. Forbes Parkhill, *The Last of the Indian Wars* (New York, 1961), 28–29, 47–48, 70–77, 116, gives the generally accepted interpretation of the last "Ute war" and Posey's death resulting from body wounds.

60. On the Myall Creek massacre, see note 59, above. On Queensland punitive raids against Aborigines, see Price's *White Settlers and Native Peoples,* 138; and Elkin's *The Australian Aborigines* 323, which states that "pacification" by force continued until the 1930s.

61. Price, *White Settlers and Native Peoples,* 138; Elkin, *The Australian Aborigines,* 323.

62. A. Barrie Pittock, *Toward a Multi-Racial Society: The 1969 James Backhouse Lecture* (Pymble, New South Wales, 1969), 5.

63. *Ibid.*, 12.

64. The arguments set forth by Macleod and Brandon and other writers are discussed in Jacobs, "British–Colonial Attitudes and Policies Toward the Indian in the American Colonies," 82 ff.

65. Francis Paul Prucha, *American Indian Policy in the Formative Years: The Indian Trade and Intercourse Acts, 1790–1834* (Cambridge, Mass., 1962), 226–227, 229; Louis De Vorsey, Jr., *The Indian Boundary in the Southern Colonies, 1763–1775* (Chapel Hill, 1966), 27 ff.

66. Nancy Oestreich Lurie, "Indian Cultural Adjustment to European Civilization," in Smith, ed., *Seventeenth-Century America, 38–60.*

67. *Ibid.*, 56–60.

68. During the summer of 1969, with a native missionary of the Seventh-Day Adventist Church who spoke fluent Pidgin English, I traveled by outboard motorboat some sixty miles beyond the Sepik River town of Ambuti to Swago village, an interior jungle settlement located on a small tributary of the Sepik. Here, with two other visiting whites, I lived for a brief period in one of the most remote, isolated native villages in Papua-New Guinea.

69. Lurie, "Indian Cultural Adjustment to European Civilization," 57.

70. *Ibid.*, 38–39.

71. Jacobs, "British-Colonial Attitudes," 86–90.

72. John D. Cross of the Mitchell Library staff in Sydney has called my attention to eighteenth-century drawings of Aborigines as exemplified by artists who were with Captain Cook. Here we have heroic, masculine figures. Later portrayals of the Aborigines, especially in the 1840s, depict them almost as monkeys with thin legs and potbellies. Examples of both kinds of drawings are in Moorehead, *The Fatal Impact*, 144–145.

73. Lurie, "Indian Cultural Adjustment to European Civilization," 38–39.

74. *Ibid.* Little Carpenter, Cherokee chief who visited England, remembered only "kind Promises" that were made to him. William L. McDowell, Jr., ed. *Documents Relating to Indian Affairs, 1754–1775: Colonial Records of South Carolina* (Columbia, S.C., 1970), 138.

75. Jacobs, "Was the Pontiac Uprising A Conspiracy?," 26–37.

76. Quoted in Weinberg, *Manifest Destiny*, 86–87.

77. *Ibid.* The relatively unknown story of the dispossesion of the Indians of the Far West during the Mexican War era is told in Robert Anthony Trennert, "The Far Western Indian Frontier and the Beginnings of the Reservation System, 1846–1851" (Ph.D. dissertation, University of California, Santa Barbara, 1969). Trennert discusses "the policy of extermination long advocated by many Texans" (p. 152) and the bitterly fought Texas Indian wars of 1846–1851 (pp. 113–154, 320–364).

78. George Simpson, Hudson's Bay Company executive, repeatedly emphasizes in his journals the importance of the Indian in the company's fur trade enterprises. He even goes so far as to consider "the effect the conversion of Indians might have on the trade," concluding that it would not be "injurious," and indeed might be "highly beneficial" if it caused the Indians to be "more industrious, more seriously . . . [concerned with] the Chase." Frederick Merk, ed., *Fur Trade and Empire, George Simpson's Journal . . . ,1824–1825* (Cambridge, Mass., 1968), 108–109. On the company's price wars, which resulted from American trading ships contacting Indians, see John S. Galbraith, *The Hudson's Bay Company As An Imperial Factor, 1821–69* (Berkeley, 1957), 138–140. A House of Commons report of 1857 strongly supported the company's desire to maintain a monopoly for the "protection of the natives against the evils of openly competitive bidding and for conservation of fur-bearing animals." Quoted in Douglas MacKay, *The Honourable Company, A History of the Hudson's Bay Company* (New York, 1936)., 274. However, as Simpson's journal demonstrates, the company's concern with profits was such

that it would not follow a policy of conservation of furbearing animals, even when the supply was being "unremittingly hunted" to exhaustion in certain areas. See Merk, ed., *Fur Trade and Empire*, 151–152.

79. H. C. Allen, *Bush and Backwoods: A Comparison of the Frontier in Australia and the United States* (Sydney, 1959), 24–25; John Wesley Powell, "From War-path to Reservation," in Wilcomb E. Washburn, ed., *The Indian and the White Man* (New York, 1964), 377–391; R. M. W. Reece, "The Aborigines and Colonial Society in New South Wales Before 1750. With Special Reference to the Period of the Gipps Administration, 1838–1846" (M.A. thesis, University of Queensland, 1969), 10–143.

80. Frederick Jackson Turner, *The Frontier in American History* (New York, 1921), 15.

81. Jacobs, "British Colonial Attitudes," 106.

82. Australian writers have perhaps been less successful in making a hero out of the "bushman" or "bushranger." See Russell Ward, *Australia* (London, 1965), 9, 58–59, 94–96. Ward develops the bushranger myth more fully in his readable volume, *The Australian Legend* (Melbourne, 1961). Romantic myths have also grown up around the picturesque Brazilian *bandeirrantes*, jungle path-finders whose movements were governed by Indian slave hunting and prospecting cycles. Richard M. Morse, ed., *The Bandeirrantes: The Historical Role of the Brazilian Pathfinders* (New York, 1965), 5, 23, 181–190.

83. Ward, *Australia*, 27.

84. All Aborigines are now citizens, and in theory possess the voting franchise and are eligible for a whole range of social service benefits. Dept. of Territories, *The Australian Aborigines,* (Sydney, 1967), 66–110. Yet the author observed that Aboriginal people are still deprived of their civil rights as in the case of the reserve at Palm Island off the coast of Queensland at Townsville. Here the administration seems to be entirely under the Queensland police and an executive officer who was formerly a plantation manager. There is a substantial body of literature on present Aboriginal problems. See, for example, *We the Australians: What is to Follow the Referendum? Proceedings of the Inter-Racial Seminar held at Townsville, December, 1967* (Townsville, 1968); Frank Stevens, *Equal Wages for Aborigines, The Background to Industrial Discrimination in the Northern Territory of Australia* (Sydney, 1968); T. G. H. Strehlow, *Assimilation Problems: The Aboriginal Viewpoint* (Adelaide, 1964); Frank Hardy, *The Unlucky Australians* (Sydney, 1968), 175–209.

85. Douglas Oliver in his perceptive, searching study of natives of the Pacific also makes this point. See *The Pacific Islands,* 425–426.

CHAPTER FOUR

THE TIP OF AN ICEBERG

Pre-Columbian Indian Demography and Some Implications for Revisionism*

Wilbur Jacobs's revisionism in Native American studies extended to the complex and controversial field of demography. In "Tip of the Iceberg," he advances, defends, and elaborates on the pre-Columbian Indian population estimates of Sherburn F. Cook, Woodrow Borah, Henry F. Dobyns, and others. Jacobs, like these scholars, argues that Native American populations in the New World were much higher than scholars had traditionally estimated. Colonial Europeans did not occupy empty wilderness inhabited by a few "primitive" Indians. Instead, they invaded regions well populated by highly developed and sophisticated Native American societies. In Jacobs's estimation, these revised demographic figures for the pre-Columbian New World should force a radical historical reassessment of the European colonization of America and raise profound questions about the nature of Native American societies.

Although controversies about the nature of the American Revolution and related topics have not diminished our interest in early American Indian-white history, we have largely ignored recent demographic stud-

**Reprinted from* The William and Mary Quarterly, *3d Ser., Vol. 31 (1974), 123–132.*

ies of the American Indian which may well give new direction to much of the rationale of colonial growth and progress. While it is difficult to make positive judgments, several scholars (mostly nonhistorians) have suggested an entirely new version of early Indian-white relations showing that Europeans had an overwhelming role in triggering an enormous depopulation of native American people.

What is involved here is truly one of the most fascinating numbers games in history, one that may well have a determining influence upon interpretive themes not only of early United States history but also of the history of all the Americas. The basic questions are these: is there evidence to show that there were some one hundred million Indians in the Western Hemisphere at the time of discovery? Further, is it true that this evidence may give us a new figure of nearly ten million Indians in the North America of 1492? And finally, is it now generally accepted by anthropologists that the figures of James Mooney, Alfred L. Kroeber, and Angel Rosenblat—some one million Indians for pre-Columbia North America and eight to fourteen million Indians for the Western Hemisphere—are now out of date? If the new estimates for native American population (allowing for disagreement among authorities but general agreement that Mooney, Kroeber, and Rosenblat are now of only relative value) are to be considered, we must now cope with new evidence that indicates there were between fifty and one hundred million Indians in possession of the New World on the day that it was "discovered." Thus we have an invasion of Europeans into areas that were even more densely settled than parts of Europe. There is even the possibility that in the late fifteenth century the Western Hemisphere may have had a greater population than Western Europe.

There is one stalwart figure who as late as 1967 continued to dispute the new evidence. This is the Latin American scholar Rosenblat. He has accused Sherburne Cook and Woodrow Borah of discarding the testimony of respectable witnesses ("el testimo de respectables testigos") in their computations to arrive at high estimates of preconquest populations in central Mexico, in Hispaniola, and the whole Western Hemisphere. In his assessments of the finds of Cook and Borah, however, he consistently avoids either a discussion of their sophisticated methodology or an evaluation of the financial records they used as a basis of calculation. Rosenblat reluctantly concludes that the use of mathematical formulas in demography has given his work the appearance of error which he sought to avoid. One of the weaker links in his argument is his hasty attempt to discredit the original estimates of Indian mortality made by Bartolomé de Las Casas. In the end, Rosenblat's long defense of the researches of Mooney and Kroeber and his manner of repeating and reprinting his own findings (in three books) to strengthen his argument are unconvincing. Furthermore, Rosenblat seems to have overlooked the findings of certain scholars who disagreed with him. For instance, he makes no analysis of the work of the geographer Karl Sapper who as early as 1924 estimated a total of thirty-seven to forty-eight million Indians for the Americas in 1492.[1]

But let us leave Rosenblat and turn to the debate as it concerned other scholars in the 1960s and 1970s. It is somewhat ironic that the growing number of writers on United States history who have successfully employed quantitative methodology should have left Indian demographic research largely to specialists in other fields such as the Berkeley physiologist Cook, and his coauthors, historian-linguist Lesley B. Simpson and Latin Americanist Borah, and anthropologists Henry F. Dobyns and Harold Driver.[2] A good place to begin probing the vital statistics in this revisionism is with Dobyns's article, "Estimating Aboriginal Indian Population."[3] What makes this study especially significant is its support of the methodology pioneered by the Berkeley scholars, Cook, Simpson, and Borah, in appraisals of Indian depopulation in California and Meso-America.[4] Dobyns reaches his hemispheric estimates by determining the demographic nadir or lowest population of Indians in regions of the Americas during the modern era when census data are available. Then he multiplies the nadir figure by a number representing the measure of Indian population loss from disease and other factors. By multiplying exponentially he estimates a population of 90,043,000 native Americans with a high projection of 112,553,750. The higher estimate results from the use of a historic depopulation ratio of 25 to 1, and the lower from a ratio of 20 to 1. Using this method, with an estimated nadir population of 490,000 aborigines in 1930 for North America, Dobyns with a ratio of 20 to 1 estimates a pre-Columbia population of 9,800,000. With the ratio of 25 to 1 he estimates a high of 12,250,000.[5]

Although Driver criticizes Dobyns's estimates by pleading for a lower nadir in estimating the aboriginal population of North America, which might result in a 50 percent cut in numbers, he nonetheless praises the new methodology.[6] Indeed, an examination of the numerous critiques published with Dobyns's estimates shows there is more agreement than dissent. What is more, the most distinguished student of Indian demography, Cook, approved Dobyns's work in conversations and in published commentary.[7] And I must confess that Dobyns in his articles and in convincing dialogue (in personal conversations and in correspondence) has persuaded me that historians may well have to adopt a whole new view of Indian demography, especially of Indian depopulation resulting from smallpox and other epidemic diseases.[8] Current literature on epidemic diseases among Indian people offers corroborative evidence to show, for example, the powerful impact of measles virus, easily airborne and devastating in its effect, upon certain native American communities, especially those in Alaska.[9]

An overriding consideration in favorably evaluating Dobyns's work is the consistently high estimates of aboriginal population in the formidable study by Cook and Borah, *Essays in Population*. These essays, concentrating on Mexico and the Caribbean, include both a brilliant analysis of methodology and a revealing commentary on historical demographic problems in making such estimates for pre-Columbian populations as the figure of twenty-five million for central Mexico. There is no question that the Cook and Borah estimates in these essays are as high as or higher than those that Dobyns provided for populations of specific areas. In some cases the Cook and Borah estimates skyrocket as high, for instance, as eight million for the pre-Columbian population of Hispaniola (present Haiti and the Dominican Republic) alone.[10] This figure far exceeds the most extravagant population estimates of contemporary Spaniards. Las Casas judged the native population of Hispaniola to be three or four million.[11]

Of course the most pressing question is how such a dense population could have supported itself in the Caribbean? On the basis of evidence given by Carl Sauer, Cook and Borah argue that the people of Hispaniola had perfected the domestication of food plants to the extent that they had a greater yield per hectare than comparable fields harvested in the Europe of 1492.[12] The supply of maize, beans, and cassava, supplemented by protein obtained by fishing and hunting, was more than enough to feed eight million people.

Given this prosperous state of affairs, there is still another obvious query to make. What brought about the sharp depopulation and finally the extermination of aborigines on Hispaniola, a process that was virtually complete by 1570? Cook and Borah, again partly relying on earlier studies by Sauer, argue that the harsh Spanish rule of native people, especially unusually brutal methods of exploiting Indian labor, was in part responsible for depopulation.[13] It was this brutality, resurrecting the Black Legend of Spanish cruelty (so distasteful to Kroeber that is may have led him to lower his population estimates of aborigines),[14] as well as disease that killed the Indians. Of all the epidemic diseases, smallpox seems to have been the scourge for millions of Indians in the Caribbean as well as in Meso-America and North America.[15]

If there were, indeed, eight million Indians on Hispaniola and some twenty-five million in central Mexico, how many were there in North America, north of the Rio Grande River? Probably at least as many as Dobyns has estimated—9,800,000 to 12,250,000—if we accept his methodology and mathematics. This estimate contrasts strikingly with the earlier ones of Mooney (1,152,950), Kroeber (1,025,950), and Rosenblat (1,000,000).[16]

Here we have it then—a new hemispheric estimate of Indian population that is almost breathtaking in its magnitude. If there ever was a tool for presentism in the writing of early American history in coping with the dispossession of the Indians, this is it. It is hard to imagine that our history can ever be the same again since we can scarcely portray the European invasion of the Western Hemisphere as the relatively quiet expansion of Europeans into sparsely settled lands. What we do with these new data and how we interpret them will be of great consequence, and we may be sure that Indian historians and the increasingly vocal American Indian Historical Society will have perceptive comments on their significance. Even if there is a general consensus that reduces the figure from one hundred million to fifty million—and some qualified investigators concede that we could hardly settle for less than that number—we must now accept the fact that the dismal story of Indian depopulation after 1492 is a demographic disaster with no known parallel in world history.[17] We must also acknowledge that the catalyst of all this was undoubtedly the European invasion of the New World.

Admittedly, what we are dealing with is the tip of a formidable iceberg. Although there is sufficient evidence to tell us that the iceberg is really there, several questions suggest themselves. There is, for instance, reason to believe that the Indians were on the verge of a population decline before the Europeans arrived. This may well have been the case in central Mexico.[18] The possibility also exists that Indian populations were already in a cycle of depopulation. Even if we accept these arguments, there remains the question of why societies so large should have been so vulnerable. Where, moreover, is there evidence of a material culture to sustain such a large population? In Meso-America the remains of great Indian societies that flourished before and after the Spanish conquest still exist despite efforts to obliterate Indian civilizations. But in North America, at the time of first contacts with the Indians, there was no concerted effort to eradicate Indian culture. Yet the material remains of prehistoric Indian societies are sparse indeed.

If we accept the evidence pointing to disease as the most important factor in causing Indian depopulation, much remains to be done in studying the ability of individual tribes to resist waves of epidemics. Another factor is the significance of such data as the physical distance between population centers. There is also the argument that the infusion of Spanish blood (as well as the blood of other Europeans) into Indian societies helped to preserve them and to strengthen them against the impact of recurring epidemic disease waves. Anthropologist Edward Spicer sets forth a cycle theory of conquest and withdrawal of Europeans who may leave behind invigorated native societies enriched by cultural exchange.[19]

Given such probabilities and possibilities, the natural reaction may well be to doubt the veracity of recent theorizing on high pre-Columbian population estimates. Since most of the new data is based upon calculations which in turn rest upon a sifting of more conventional evidence, the hard documentary sources are lacking. Yet because such documentation is missing we may well be in error if we assume that the new methodologies cannot be trusted. In Indian history, as in the histories of other minorities, we are finding that the conclusive evidence found in Anglo-Saxon sources is often impossible to obtain.

My examination of the work of Cook, Borah, Sauer, Dobyns, and their critics leads me to believe that they have discovered a great historical iceberg concerning Indian populations. My own investigations, partly based upon examination of modern medical literature concerning Indians[20] and upon field research, tend to bear out their findings, as does archival evidence.

For instance, in my study of native-white contacts in Australia I have found that Australian Aborigines (who have a marked resemblance to certain Indian tribes of the American Southwest) had a swift depopulation after first contacts with whites in the 1830s. English agents, called "Protectors," reported a great "mortality" caused by disease and forced transfer to unfamiliar surroundings.[21] In my field trips to modern reserves of the Aborigines in Darwin, Alice Springs, and at Palm Island (off the coast of Queensland), there was a problem of being admitted at times because of recurring quarantines. Everywhere on reserves, despite medical precautions, native people seemed to be fighting off one epidemic after another of typhoid and other diseases. Leprosy, although partially checked, still persists among the native people of Queensland. Despite these onslaughts of disease, the Aboriginal population is now increasing, largely because of the Australian government's excellent social programs and medical facilities. The overall evidence, however, shows that widespread depopulation took place in early Australian history and that some native people, such as the Tasmanians, were finally wiped off the face of the earth.[22] Considering these facts, it is not surprising that Australia's leading anthropologist, the late A.P. Elkin, saw many parallels in the history of native-white relations in his country and in North America.[23]

In my field work and archival research on the impact of European contact with the native people of Papua, New Guinea, I found that the Melanesians withstood European invasion with more success than either the Australian Aborigines or the American Indians.[24] And the Polynesians, especially those of the Hawaiian Islands that I have studied, also seem to have withstood the disease and the cultural impact of Europeans better than the Indians or the Australian Aborigines. Nevertheless, Anglo-American pressure to dispossess Hawaiian native people of their lands and to stamp out their religion was (and continues to be) unrelenting.[25]

I found additional data bearing out the findings of Cook, Borah, and Dobyns in my field work and examination of archival data relating to the Arawaks, who were all but exterminated on Jamaica, Hispaniola, and other Caribbean Islands in the sixteenth century. Examination of the artifacts at the Arawak Museum and Burial Ground, maintained by the Institute of Jamaica, indicates that these peaceful, friendly people were skilled fishermen. It is certain that they ate large amounts of shellfish which, in addition to their skills in gardening, could have enabled them to feed a large population. One big Jamaican village site has five thick layers of shells, pottery, and bones dating back to about 1000 A.D. There are many other such sites, especially along the Jamaican coastline, formerly occupied by Arawaks.[26]

Although Cook and Borah find that the Arawaks were gone by about 1570 in Hispaniola, a handful of them survived in Jamaica as late as 1598. Spanish governmental manuscript material, preserved at the Institute of Jamaica, reveals that in 1598 the government attempted to protect the scattered survivors "from the many that there were" by giving them an independent village sanctuary. This effort, however, was bitterly resisted by the local Spanish ranchers who succeeded in keeping the Indians in virtual slavery as fieldhands or cattleherders.[27] Later English records, some as late as 1700, show Indians as slaves or servants on Jamaican plantations, but there is no way to determine if these Indians were surviving Arawaks or other tribesmen.[28] Many Indians, after seventeenth-century wars in the British North American colonies, were sent as slaves to the British West Indies.

Thus, although the evidence is fragmentary and scattered among a number of different sources, it does show that native peoples have greatly suffered under the impact of the European invasion. In some cases, as clearly illustrated by the experience of Australian Aborigines and the Arawaks of the Caribbean, there was a sharp decline in native population after first contacts with whites, even the wiping out of indigenous native communities that accompanied dispossession and seizure of their homelands.[29] But a still larger transformation took place in the Western Hemisphere in the great demographic disaster involving the disappearance of millions of Indians following the first invasions of Europeans. The dimensions of this disaster have now been outlined by Cook, Borah, and Dobyns. It does indeed appear to be the tip of an iceberg of tremendous proportions.

NOTES

1. Ángel Rosenblat's argument rest on the figures of James Mooney and A.L. Kroeber. Mooney, *The Aboriginal Population of America North of Mexico,* Smithsonian Miscellaneous Collections, LXXX (Washington, D.C., 1928), 33,

estimates the total pre-Columbia Indian population north of Mexico including Greenland at 1,152,950. Kroeber, *Cultural and Natural Areas of Native North America,* University of California Publications in American Archaeology and Ethnology, XXXVIII (Berkeley, 1939), 131, 166, reduced Mooney's estimate to 1,025,950, nearly 10%, and made a hemispheric estimate of 8,400,000. Rosenblatt's hemispheric pre-Columbian estimate is 13,385,000, but his estimate for North America, north of the Rio Grande, is slightly lower than Kroeber's, an even 1,000,000. His population tables are found in his *La Población Indígena de América desde 1492 hasta la Actualidad* (Buenos Aires, 1945), 92, and in a revised work with unchanged pre-Columbian estimates, *La Población Indígena y el Mestizaje en América I: La Población Indígena, 1492–1950* (Buenos Aires, 1954), 102. Rosenblat defends his estimates in a third book, *La Población de América en 1492: viejos y nuevos cálculos* (Mexico, D.F., 1967), especially 1–9, 11–16, 81. Karl Sapper's statistical tables are printed in Julian H. Steward, ed., *Handbook of South American Indians,* V: *The Comparative Ethnology of South American Indians,* Bureau of American Ethnology Bulletin 143 (Washington, D.C., 1949), 656.

2. The key publications of these scholars on the Indian demography debate are in the footnotes which follow.

3. Henry F. Dobyns, "Estimating Aboriginal American Population: An Appraisal of Techniques with a New Hemispheric Estimate," *Current Anthropology,* VII (1966), 395–416.

4. *Ibid.* 403–407. Among the published works of the authors Dobyns relies primarily upon S.F. Cook, *The Extent and Significance of Disease among the Indians of Baja California, 1697–1773,* Ibero-Americana, 2 (Berkeley, 1937), 9–14; Sherburne F. Cook and Lesley Byrd Simpson, *The Population of Central Mexico in the Sixteenth Century,* Ibero-Americana, 31 (Berkeley, 1948), 19–22; Sherburne Cook and Woodwrow Borah, *The Aboriginal Population of Central Mexico on the Eve of the Spanish Conquest,* Ibero-Americana, 45 (Berkeley, 1963), 6, 22–44, 157. Dobyns stresses that Cook's researches reveal that a "fatal defect" in both Kroeber's data and method was his ignoring of disease, especially epidemics: "Estimating Aboroginal American populatin," *Current Anthropology,* VII (1966), 411.

5. Dobyns, "Estimating Aboriginal American Population," *Current Anthropology,* VII (1966), 415. This estimate is presumably for Indians north of the Rio Grande, but Dobyns is not entirely clear on this point. Dobyns's final table of estimated populations is reprinted in Virgil J. Vogle, ed., *This Country Was Ours: A Documentary History of the American Indian* (New York, 1972), 253. As early as 1962 Borah estimated a hemispheric pre-Columbia population of

100,000,000. Dobyns states that Borah's depopulation ratio would fall between 20 and 25 and that Borah's estimate for population density per square kilometer implies 2.4 persons. Dobyns estimates 2.1 density: "Estimating Aboriginal American Population," *Current Anthropology*, VII (1966), 414. He cites Woodrow Borah, "¿América como modelo? El impacto demográfico de la expansión europa sobre el mundo no europeo," *Cuadernos Americanos,* VI (1962), 176–185.

6. Driver's commentary is published with other critiques of Dobyns's article in *Current Anthropology*, VII (1966), 429–430. He questions Dobyns's exact geographical definition of North America and argues that his nadir population figure should be lowered (250,000 in 1890 for the area of the United States) and that his overall estimates might thus be cut by as much as 50%. Driver is also critical of Dobyns's estimate of 30,000,000 for the population of central Mexico which in turn was based upon the 1963 estimates of Cook and Borah ranging from 25,000,000 to 30,000,000. In their latest study, Cook and Borah arrive at the figure of 27,650,000. See Sherburne F. Cook and Woodrow Borah, *Essays in Population History: Mexico and the Caribbean*, I (Berkeley, 1972), 115. See also Harold E. Driver, *Indians of North America,* 2d ed. rev. (Chicago, 1969), 63–65, and his "On the Population Nadir of Indians in the United States," *Current Anthropology*, IX (1968), 330.

7. Cook's published comments on Dobyns's estimates are in *Current Anthropology*, VII (1966), 427–429.

8. Besides Cook's work, cited in n. 4, see Henry F. Dobyns, "An Outline of Andean Epidemic History to 1720," *Bulletin of the History of Medicine,* XXXVII (1963), 493–515, and Wilbur R. Jacobs, *Dispossessing the American Indian: Indians and Whites on the Colonial Frontier* (New York, 1972), 136, 136n, 191n, 214n. For the North American Indian population control, see *ibid.*, 130, 130n, 163.

9. I am indebted to John C. Bolton, M.D., of San Franciso, a specialist in the study of modern epidemics, who has introduced me to the vast literature on the subject. Jacob A. Brody, M.D., of Alaska and New York State, and his associates have written several key articles on recent measles epidemics among Eskimos, Alcuts, and Alaskan Indians that demonstrate the severity of measles infection. See, e.g., "Measles Vaccine Field Trials in Alaska," *Journal of the American Medical Association*, CLXXXIX (Aug. 3, 1964), 339–342. Disease patterns of Indians in the southwestern parts of the United States where there have long been close association and intermarriage with non-Indians reveal no

special weakness for epidemic diseases but that Indians, like their white neighbors, suffer from diabetes, tuberculosis, and cancer, although "hypertension is apparently less frequent than among white persons." See Maurice L. Sievers, "Disease Patterns among Southwestern Indians,: *Public Health Reports,XI (1966), 1075–1083 (quotation, p. 1082). See also S.M. Weaver, "Smallpox or Chickenpox: An Iroquoian Community's Reaction to Crisis, 1901–1902,"* Ethnohisotry, XVIII (1971), 361–378, and Mark A. Barrow *et al. Health and Disease of American Indians North of Mexico: A Bibliography, 1800–1969* (Gainsville, Fla., 1972), 57–58. Cook's current research on disease and the New England Indians indicates that earlier population estimates for the North Atlantic littoral (by Mooney and others) are too low. Correspondence in 1973 between W.R. Jacobs and Sherburne Cook.

10. Cook and Borah, *Essays in Population,* 407.

11. A table of Las Casas's estimates in Dobyns, "Estimating Aboriginal American Population," *Current Anthropology,* VII (1966), 397. Philip Wayne Powell, *Tree of Hate: Propaganda and Prejudices Affecting United States Relations with the Hispanic World* (New York, 1971), discusses the significance of Las Casas's indictment of Spanish brutality and the origins of the Black Legend of Spanish cruelty. He argues, 139–159, that the causes of Indian depopulation were very complex.

12. Cook and Borah, *Essays in Population,* 408, citing Carl Sauer, *The Early Spanish Main* (Berkeley and Los Angeles, 1966), 67–69, 157, *et passim.* Sauer, in his recent volume, *Sixteenth Century North America: The Land and the People as Seen by Europeans* (Berkeley and Los Angeles, 1971), 58, 59, 71, 205, 252–253, 286–288, 294–295, stresses the agricultural skills of the North American Indians, as does Wilbur R. Jacobs, "The Indian and the Frontier in American History: A Need for Revision," *Western Historical Quarterly,* IV (1973), 50–56.

13. Cook and Borah, *Essays in Population,* 409, and Sauer, *Early Spanish Main,* 202–204, 283–289. Cook and Borah, however, still regard disease as the most important cause of Indian depopulation. Letter from Woodrow Borah to W.R. Jacobs, Apr. 13, 1973.

14. Dobyns, "Estimating Aboriginal American Population," *Current Anthropology,* VII (1966), 397, alludes to this point which seems to have validity if one notes the absence of condemnation of Spanish brutality in Kroeber's writings dealing with the Spanish occupation of the New World.

15. Dobyns, "Estimating Aboriginal American Population," *Current Anthropology,* VII (1966), 510–412.

16. See n. I.

17. Magnus Mörner in his perceptive study, *Race Mixture in the History of Latin America* (Boston, 1967), 50, was one of the first historians of Hispanic America to give tentative acceptance to the figure of 50,000,000, and about the same time J.H. Parry, *The Spanish Seaborne Empire* (New York, 1966), 213–228 (a chapter entitled "Demographic Catastrophe"), accepted the figure of 25,000,000 for the preconquest population of New Spain. Historians of the United States, however, seem to have overlooked the increasing volume of literature on Indian demography, and anthropologists who had a role in the debate have moved on to what appear to be other controversies in demography, in nonhistorical topics such as fecundity and recent trend analysis.

18. This point is mentioned in the debate over the 25,000,000 estimate for central Mexico. But the key issue is the reliability of evidence behind the estimate, for once the 25,000,000 figure for central Mexico is accepted, other high estimates based upon exponential calculations follow. For Rosenblat's criticism of Cook and Borah on this estimate see his *La Polación de América en 1492,* 78–81.

19. Edward H. Spicer, *Cycles of Conquest: The Impact of Spain, Mexico, and the United States on the Indians of the Southwest, 1533–1960* (Tucson, Ariz., 1962), 568. He neglects, however, to point out that examples of the withdrawal of Europeans are few indeed.

20. See examples in n. 9.

21. Protector George Augustus Robinson's reports and letters of the 1830s detail the depopulation of the Aborigines. See especially his "Reports on the Tasmanian Aborigines," A 512, 70ff, and his letter to Gov. George Arthur, Sept. 9, 1839, Mitchell Library, Sydney, Australia.

22. Details of this unpleasant story are in Wilbur R. Jacobs, "The Fatal Confrontation: Early Native-White Relations on the Frontiers of Australia, New Guinea, and America—A Comparative Study," *Pacific Historical Review,* LX (1971), 293–309.

23. See Elkin's classic study, *The Australian Aborigines,* 3d ed. (Sydney, 1954), 29, 44, 156–162 *et passim*, for his strictures on missions and missionaries among Aborigines and California Indians.

24. The complexities of this contest are discussed in Jacobs, "The Fatal Confrontation," *Pac. Hist. Rev.*, LX (1971), 203–309.

25. On dispossession of native Hawaiians from their lands see, e.g., translations of "Native Testimony," I: Land of Papua, in Gov. Kekuanaoa's sworn

testimony, Mar. 18, 1846, Hawaiian Archives, Honolulu. The "Thaddeus Journal," 1819–1820, Hawaiian Mission Children's Society, Honolulu, has entries through March and April 1820 which reveal the almost astonishing fear and hostility New England missionaries had toward native religions. Modern American Mormon missionaries are among the most enthusiastic proselytizers among native Hawaiian people. Their success is assured by the popularity of their operation of a Polynesian Cultural Center on Oahu.

26. I am indebted to Professor Richard S. Dunn of the University of Pennsylvania for assistance in locating this Arawak burial site and for other help in carrying out my investigations in Jamaica.

27. Translations from the Archivo De Indias, Seville, AGI, 54–3–28, Dec. 26, 1598, Audiencia de Santa Domingo Isla de Jamaica, Pietrsz Bequest, Institute of Jamaica, MST 29, Vol. 2. This manuscript volume also contains documents concerning the unsuccessful governmental project to create a special Indian village sanctuary.

28. See the English estate inventories preserved at the Jamaica Archives, Spanish Town, Jamaica. The following inventories are representative of those listing Indian slaves: Inventory, Book 5, 48B; Inventory of James Pinnock Junior, enrolled at Mar. 22, 1700, *ibid.*; Inventory of Elizabeth Potts, Widow [Mar. 1700], 67, 67B, *ibid.*; Inventory of chattels and debts of Thomas Harry [Jamquier], [Mar. 1675], Inventory Book I, 1674–1675. The latter inventory is on microfilm.

29. Jacobs, *Dispossessing the American Indian,* 19–30, 126–172, discusses the significance of interrelationships between dispossesion and depopulation of native peoples. See also Douglas L. Oliver's learned study, *The Pacific Islands,* rev. ed. (New York, 1961), 1–80.

CHAPTER FIVE

THE INDIAN AND THE FRONTIER IN AMERICAN HISTORY— A NEED FOR REVISION*

Wilbur Jacobs is critical of history told only from the Anglo point of view. In this influential article published in 1973, he argues that traditional American history should include perspectives on ethnic minorities. A genuinely pluralistic history would stimulate fruitful scholarly debate and long-needed revision and would create a more accurate portrait of the American past. Focusing on Native American history, Jacobs demonstrates how the Indian point of view would force historians to reconsider accepted interpretations of frontier history. The essay explores four areas: pre-Columbian Native American population levels and postcontact population decline, the North American environment and Native American conservation practices, and the influence of Indian culture on frontier mythology.

The proposition that I am setting forth here is that although much traditional Indian-white frontier history is of great significance in illuminating our knowledge of the past, there is a need to focus more of our attention on what might be called the Indian point of view. In fact, I am suggesting that we revise certain of our traditional ideas about the fron-

*The Western Historical Quarterly *4 (January 1973), 43–56.*

tier in American history with a hope of seeking a balance to offset some of the widely accepted interpretations that have repeatedly appeared in our textbooks and in many learned journals. If we are going to tell the whole story of Indian-white relations, we must make an all-out attempt to picture the clash of cultures so that there will be an understanding of both cultures, not just one. Thus, to give more attention to the Indian side is not necessarily to plead for the Indian point of view. What we are seeking, it seems to me, is a wider basis of truth, a better understanding of what has happened in the past. We may then have a truer account of what actually happened to both Indians and whites and of the emergence of a new culture that formed on the frontier at the point of contact.

In short, we need to ask ourselves if the history of the United States is only the history of white people, or if it also includes the American Indian, as well as other ethnic minorities. Just because the documents are missing, as Bernard DeVoto once told me, we have assumed that in some cases there was no history at all. In fact, by basing our research on documentary evidence, American historians, including specialists in frontier history, have often tended to assume that the history of America is largely a function of white culture alone. My purpose here is to refute that assumption; indeed, the Turnerian theme of progress and development as an explanation of frontier advance is largely an interpretation of Euro-American white history of political party changes and the evolution of sectional characteristics and national ideals. It has little to do with Indians, blacks, Asians, Mexican Americans, or other minorities that make up the bulk of the population in large American cities, including Los Angeles, Chicago, and New York.

But let us turn to the subject of the Indian's past in United States history to see how significant it really is. What follows is an examination of four representative areas of Indian-white history where evidence suggests revision is needed. In fact, we may need to turn some of our established views upside down and put them through a wringer if we hope to comprehend what has actually happened. For example, I will examine some misconceptions about Indian population and suggest revision of the "conquest of the West" theme in American frontier history, which has provided the rationale for dissipating our resources and for relegating the Indian to historical insignificance. I will also describe several important Indian concepts on conservation, and finally I hope to show at least one aspect of the significance of the Indian that has been overlooked in our development of American culture.

Much of what I have to say under these headings comes from sources not always used by historians, such as records of Indian speeches and treaties, Indian oral history tapes, the great mass of data compiled by our leading anthropologists and geographers, plus insights from the writings of modern Indians, such as N. Scott Momaday and Vine Deloria. I will not attempt to question all the myths about Indian-white history in America's past, but rather to hold some of them up for critical examination.

Let us begin by looking at this matter of population and population growth. Frederick Jackson Turner's whole concept of American frontier history is founded upon white census data. He based his famous essay of 1893, describing the significance of the frontier in American history, on the 1890 census count, which indicated that the white frontier of settlement had come to a halt. Likewise, in twentieth-century history, population pressures have been very important in recent world history. We might even say overpopulation was a direct cause for the outbreak of World War II because Germany, Italy, and Japan argued they needed more territory for their oversupply of people.

What about Indians? Why are they a force in the history of human population? Until about ten years ago we generally accepted the data set forth by James Mooney and A. L. Kroeber, leading anthropologists of the last generation, that in 1492 there were approximately one million Indians in North America, give or take a few hundred thousand. In all of the Western Hemisphere Kroeber estimated a total of 8,400,000 Indians at the time of discovery. One of his later contemporaries concluded after careful evaluation of the evidence that there might be as many as 13,385,000.

But in the 1960s and early 1970s, as a result of the statistical findings of Henry Dobyns, Woodrow Borah, Shelburne Cook, Harold Driver, and other scholars, our whole concept of Indian population and methodology in computing estimates has changed. For instance, Dobyns, an anthropologist who now teaches at Prescott College, has set forth what is becoming a recognized projection of 90,043,000 Indians at the time of discovery for the Western Hemisphere and 9,800,000 for North

America. Cook and Borah, combining their skills as physiologist and historian in a 1971 study on native populations of Mexico and the Caribbean, provide figures that are even higher than Dobyns's, for instance their estimate of 8,000,000 for the pre-Columbian island of Española (modern Haiti and Santo Domingo). Using the same exponential mathematical techniques as Borah and Cook, Dobyns estimates that the figure for the whole Western Hemisphere might be over one hundred million.

If Dobyns is correct, this number is probably more than the total population of Western Europe at the time Columbus discovered America! How does all this change our history? Indeed, we may ask ourselves if our history could ever be the same if we historians accept such evidence as scholars in other fields have done. Certainly we cannot ignore the mass of data, the new methodology or the rigorous evaluation of factual data behind such estimates.

We must now acknowledge that the discovery and settlement of the New World cannot ever be described as the movement of Europeans into areas sparsely occupied by native peoples. No longer can commencement speeches about the peaceful occupation of a largely vacant land be accepted with credibility. In fact, what we are dealing with here seems to be a genuine invasion of the New World by Europeans and the dispossession of a hundred million or more native indigenous peoples. The whole rationale of colonial and frontier history in the Western Hemisphere and the arguments explaining the origins of New World nationalities and nations must somehow cope with this new mass of evidence.

What happened to all those Indians? Cook and Dobyns, researchers in the spread of epidemic diseases among Indians, argue convincingly that millions of Indians were killed off by catastrophic disease frontiers in the form of epidemics of smallpox, bubonic plague, typhus, influenza, malaria, measles, yellow fever, and other diseases.[1] (Besides bringing Old World strains of virus and bacteria, Europeans brought weeds, plants, rats, insects, domestic animals, liquor, and a new technology to alter Indian life and the ecological balance wheel.) Smallpox, caused by an air-borne virus, was and is about the most deadly of the contagious diseases. Virulent strains, transmitted by air, by clothing, blankets, or by slight contact (even by an immune individual), snuffed out whole tribes, often leaving only a handful of survivors. Although some kinds of epidemic diseases might be reduced to a mild virulence among Indians (as among whites) after generations of exposure, smallpox was undoubtedly the Indians' worst killer because it returned time and again to attack surviving generations of Indians to kill them off too. Where did it all start?

Dobyns is convinced, and he has convinced me too (in conversations, correspondence, and articles), that the epidemic disease frontiers began with the first European contacts with the New World in the 1400s and 1500s by explorers, fishermen, and the first colonizers. There is evidence, for example, in Carl Sauer's book, *Seventeenth-Century North America, The Land and the People as Seen by the Europeans,* that De Soto's expedition of 1539–1542 spread diseases over a wide area, greatly reducing Indian population. There were, then, disease epidemics after the discovery of America that swept like wildfire through the whole Indian world, swiftly moving over the entire continent before Jamestown and Plymouth were going concerns. Thus, in American frontier history we have a more significant frontier to discuss and think about, a virulent disease frontier that preceded the often destructive impacts of fur trading, mining, and agricultural frontiers, which also had much to do with the continued spread of epidemic diseases and the depopulation of the Indian.

Before leaving this subject of Indian population, we should make one more significant observation. Experts have determined that by the early 1900s the North American Indian reached the nadir in his numbers with the lowest population in his history. Since colonial times the basic Anglo-American rationale for dispossessing the Indian has been that his way of life was passing, his numbers were decreasing, and it was inevitable that the white man's technically superior society, a Christian society based upon use of the tilling of the land in accordance with the wishes of "the Creator," would eventually bless the entire continent with its beneficial civilizing influences. This injunction about use of the land, especially with the idea of using the "Creator" to justify dispossessing the wild Indian, was a favorite of political leaders, such as William Henry Harrison and Thomas Hart Benton, who seemed to envision the eventual disappearance of the Indian altogether.

But what has since happened? After the 1920s the Indian population began to rise so that projections of the 1970 census reveal the existence of a million or more native American Indians, all of them citizens of the United States. Well, this may give us pause for thought. If the twin devils from the white man—disease and liquor—were no longer causing Indian depopulation, if, in fact, he were actually becoming a political force with articulate spokesmen on his own side, what about this myth of taking the land away from a dying race? Indeed, the descendants of those hundreds of thousands of Indians who were dispossessed by our pious ancestors of the nineteenth century are still with us. It was one thing to take land from a people who were tragically dying, but it is something else to deal with modern Indians who, in increasing numbers, claim they were robbed and cheated by the United States government over a long period of years. Thus, the population curve has made a full cycle. About a million Indians (especially those at the American Indian Historical Society who publish *The Indian Historian*) will demand a rewriting of our history. In fact, the Indian Historical Society of San Francisco has already published a book showing the multiplicity of errors and the racism in secondary history textbooks. In its next critique the society takes on college textbooks, which appear from initial findings to have an even greater margin of error.

What can we further conclude from our comments on population figures and the American Indian? An inescapable answer, I think, is that the sheer presence of a million American Indian citizens (many of them angry, articulate, and militant) is going to bring about a wholesale change in much of our colonial- and frontier-history interpretations.

With that in mind, let us leave the controversy about population statistics and the Indian and talk about a favorite grievance of modern Indian writers—the white man's historic exploitation of the soil, the turning of a great pristine wilderness into a foul-smelling, poisoned, soil-exhausted land. Is there truth in this charge that our dollar-chasing civilization has all but ruined America and that developers even now are making every effort to despoil our last wilderness sanctuaries by oil pipeline projects, clear-cutting forests, strip-mining, open-pit mining, and widespread use of dangerous pesticides and poisons in agriculture and in sheep- and cattle-raising country? If there is any doubt about this, for those who are concerned about such questions, you need only turn to the *Sierra Club Bulletin* and the Audubon Society's magazine of the past two years, or examine the files of leading newspapers, such as the *New York Times* and the *Los Angeles Times*.

In such publications there is the tacit assumption that our great ecological crisis is today's crisis, that it is largely a contemporary dilemma. But this is not necessarily so, according to those who have studied the problem. The historian's interest in tracing the destruction of our resources was sparked by the 1972 annual meeting of the Organization of American Historians in Washington, when two hours were given to examining evidence of dismal, tragic dust-bowl history as recorded by Woody Guthrie's ballads and Pare Lorentz's great film, "The Plow That Broke the Plains." What a tale of shortsighted wastefulness they tell. Despite the Frederick Jackson Turner myth about the conquest of the West by waves of fur traders, miners, and farmers, most of it portrayed in the technicolor of progress and development, it is becoming increasingly evident to many of us that generations of pioneers consciously and unconsciously contributed to the destruction and depletion of our vanishing resources.

What actually happened in our frontier history of land occupation is perhaps best measured, not by a historian, but by the eminent geographer Carl O. Sauer in a chapter called "Plant and Animal Life Destruction in Economic History," in his book of essays called *Land and Life*. Sauer argues that "the modern world has been built on progressive using up of its real capital." Surprisingly, Sauer points out that one of the "worn out" parts of the world is the United States, an area of relatively recent settlement in world history. "The United States," he writes, actually "heads the list in exploited and dissipated land wealth. Physically, Latin America is in much better shape than our own country. The contrast in condition of surface, soil, and vegetation is apparent at the international border between the United States and Mexico." Sauer, after making this point that America, erroneously considered the wealthiest nation in the world, is actually a country impoverished because its land has been looted, goes on to give doubtful readers proof. If you really want to see what the land was like before it was depleted, go to Mexico, he says. Indeed, for Americans who want to see for themselves "a reconstruction of soil profiles, and normal vegetation of California," Sauer says, "we must go to lower California." And "Chihuahua," he argues, "shows us what New Mexico was like a generation ago." Think about that. Think of the enormous forces at work over hundreds of years that would bring about such a dissipation of the land and the frontiers of the fur traders, grazers, miners, farmers, and industry.

Certainly Sauer's message, first written in an article of 1938, has real significance for us today. He is telling us that basically, as far as the land in the Southwest is concerned, we are a poor country. What Sauer told us in 1938 is now affirmed by a study on the limits of American growth by a team of Massachusetts Institute of Technology specialists. By the year 2100 our base in resources will have been so depleted, they say, that a worldwide production could only support a living standard of the nineteenth century. According to this MIT study (based upon mathematical formulas spanning the past and present to project into the future), economic growth, population, and industrial expansion increase in geometric ratios to cause similar geometric expansion of pollution and resource exhaustion. Thus, like compound interest (demonstrated by the increase in $24 given Indians for Manhattan Island in the 1600s, which would be worth $25 billion in 1972), our economic growth causes geometric or exponential resource exhaustion and pollution. Yet the irony of it all is that the Reverend Thomas Malthus made some similar predictions. Without the computer research, Malthus told us that population increased exponentially and food supply only increased at a constant ratio. Disease, war, and starvation, he sadly predicted, would step in to keep the balance. While not in the style of Malthus, Indian leaders have made similar prophesies.

Do we need to sound an alarm? Many American Indians think so despite their bad judgment in signing ninety-nine year leases to bring power companies to their reservations, as at the "four corners," where most conservationists believe an ecological calamity is now in the making. And in southern California, Riverside's mayor pleads that his community is subject to a nightmare of polluting smog. Over 260 days a year this unfortunate city has air pollution in the danger zone. Yet it is fascinating to note that southern California just recently began piping in water from the northern part of the state, which will be used for expanding the industrial and population base of this heavily and dangerously polluted area.

The early American Indians did not know about Riverside's dilemma, computer research, the findings of Thomas Malthus, or the warnings of Carl Sauer; but somehow they solved the ecological dilemmas that all Americans now face. The Indian, who was basically a conservator, whose religious and totemic beliefs and tribal customs prevented him from following a policy of soil exhaustion or animal extermination, is all but ignored in our history except to account for various wars, uprisings, and conspiracies, which were put down by the gun and the sword. In all, the Indian seems to have been regarded only as a kind of geographical obstacle to the westward movement of white people, as a kind of difficult mountain range or pass through which swarms of whites had to pass on their way to the Midwest, to Utah, Oregon, and California. Turner dismissed the Indian in his lectures and writings as a mere "consolidating influence," a physical danger on the frontier that forced the settlers to consolidate for mutual defense and set up pioneer forms of government.

Despite the tendency to relegate the Indian to historical insignificance, he may have something to say of importance for all Americans, especially on the subject of ecology. Here, for example, are some facts about the Indian that show how completely different his attitude was toward the land and his natural surroundings. These illustrations are largely about the woodland Indians, those great tribes that once occupied all of America's forest lands east of the Mississippi River.

First. It is reasonably certain that the Iroquois, as well as many of the southern woodland tribes, had a form of birth control to prevent overpopulation and exhaustion of food supply. Both John Lawson and Benjamin Franklin wrote about the manner in which Indian families were able to practice what we now call a type of family planning.

Second. In regard to the soil, many of the woodland tribes, including the Shawnee of Tecumseh's tribe, believed that the soil of Mother Earth was so sacred that it should not be defiled by metal tools. Mother Earth's body could only be caressed with a hoe made from the shoulder blade of an animal or a pointed stick. In contrast, think of our gigantic earth-moving machines that rip whole mountains away for dams and freeways, few of which we really need anymore, but nevertheless continue to build. Frank Speck, eminent anthropologist of the woodland Indians, pointed out that some of these great farming people of the woodland tribes resisted the use of metal plows for farming in the nineteenth century when they lived on Oklahoma reservations. Even then they refused to disturb the soil with metal tools, Indeed, the Indian's religion in many respects almost precluded carrying on the kind of agriculture the white man associated with Christianity.

Third. The Hurons and Iroquois (as well as many other woodland farming tribes) were able to keep extensive cornfields productive for a dozen years or more without resorting to what modern farmers depend upon—metal tools, fertilizers, and pesticides. Their fields were burned in the fall and in the following spring were planted with both maize and beans. Burning released nutrients for the soil, but there is no question that the growing of beans and maize together increased the soil's yield. Today we know of nitrogen fixation by bacteria to roots of leguminous plants, but with the Iroquois it was a belief in the spiritual union of beans and corn that gave vitality to the soil.

Fourth. Many of the woodland Indian tribes maintained elaborately planned and well-marked family hunting territories where, for example, beaver population was carefully watched and nurtured to prevent the extermination of these valuable animals. However, it is not generally known that the later introduction of the fur trade, especially the fierce competition between the French and the Albany traders for colonial beaver, with enticements of rum, guns, hardware tools, and clothing, actually brought about an almost complete extermination of both beaver and otter in the Iroquois country by the year 1640, only thirty years after the coming of the white man to colonial New York. The impact of the trading frontier so shattered native concepts of conservation that the Iroquois pursued the beaver into the Great Lakes country in the middle of the seventeenth century. It was on the fringes of their territory that the Iroquois hunters fought off Ottawa, Illinois, Twightwee (Miami Indians of Ohio), and other Algonquian tribes. By the 1680s the Iroquois were fighting for their share of distant Great Lakes beaver-hunting territories. As the colonial historian Cadwallader Colden wrote: "The Five Nations have few or no Bever in their Country, and for that Reason are obliged to hunt at a great Distance, which often occasions Disputes with their Neighbours about the Property of Bever." Clearly, this beaver war actually developed into one of the great bloodlettings of American history by the 1680s, as the Iroquois attempted to destroy not only Canadian Indian rivals in the beaver trade, but also their French allies.

The significance of this saga about beaver hunting and colonial warfare is that it helps to silhouette forces that propelled the early frontier to the West. It also illustrates the fact that although the Indians were basically conservationist (and the Iroquois easily lived within the ecological balance of the northeastern forest before the decimation of forest wildlife), unfortunately they were partners in the destruction of beaver in their eager search for furs to obtain guns, liquor, and other articles the Albany traders had on supply. Another factor was of equal importance—the ecological catastrophe of the that decimation of the beaver in the fur trade, a valuable animal so easily provides a habitat in ponds for many other animals and birds as well as for fish and a variety of plants.

Surprisingly, the story of the extermination of beaver is almost always treated in our histories as fur-trade enterprise, a step in the progressive development of American society, and the taming of the West. Reflecting the frontiersman's viewpoint, white historians have generally equated frontier progress with fur trade business because progress meant frontier expansion and implied elimination of the Indian frontier, a barrier to peaceful white occupation of the continent. What has seldom been pointed out is that the disappearance of the beaver closely followed the disappearance of the Indian as free-roaming tribesmen. By 1840, two hundred years after they had been cleaned out of the area of New York state, the beaver, according to Kit Carson, had been practically killed off in the Rocky Mountain area. It was in this decade according to a recent doctoral dissertation, that the Indian reservation system in the Far West had its beginning.[2] The beaver, along with the plains buffalo, the antelope, the California grizzly, and the free, independent Indian, all gave way to the white frontier of conquest about the same time.

Although Indians themselves do not always practice what they preach as far as ecology is concerned, it is nevertheless true that Indian ideas of conservation and reverence for the land are part of our heritage. Our Judeo-Christian ethic, which has tended to cast Indian society as *evil* and *pagan* and white civilization as *good* and *progressive,* has also tended to obscure what the Indian's conservation message really is. We may now come to understand, as an eminent historian of science has argued, "to a Christian a tree can be no more than a physical fact. The whole concept of the sacred grove is alien to Christianity and to the ethos of the West." We may also come to understand that although whites have seen their sacred spirits in the form of a trinity and not like the Indian in sacred plants, mountains, trees, and animals, both whites and Indians may eventually agree on a form of ecological righteousness to cope with the disaster threatening to overwhelm us. Certainly the Indian concept of a reverence for Mother Earth may help us to understand that there may be a religious side to conservation as well as a practical outlook.

Surprisingly, the Indian may also have something to do with our mythical heroes in western history, especially the frontiersman and mountain man as symbolized by Manuel Lisa, Jedediah Smith, Kit Carson, and Davy Crockett. It seems to me that these individuals are historical heroes because, in a sense, they combine many of the traits of both Indians and whites. Herein is a key to understanding part of the powerful influence the Indian has had upon American society. Our literary heroes in Cooper's *Leatherstocking Tales* are often an Indian-white combination. For example, the guide, the scout, as he is seen in the *Deerslayer* is a character of heroic mold representing characteristic virtues of both the Indian and the pioneer. Similarly, Robert Rogers, sterling frontiersman hero of Francis Parkman's *History* and Kenneth Robert's novel *Northwest Passage,* is likewise a woodsman par excellence, respectful of nature, and a woodsman-fighter who almost outdoes the Indians in the arts of camouflage, surprise attack, and survival in the winter wilderness. It is out of this tradition that America's greatest hero figures have come—the Indian guide, the scout, ranger, mountain man, plainsman, and, finally, the cowboy. These archetypes emerge as our most imitated and admired historic characters to become heroes in history, literature, and drama. More often than not such figures are more Indian than white in their life-style.

And in their own way, the heroic Indian leaders of the past are gradually assuming more stature if only as noble red men of American history. Massassoit, great Wampanoag chief, who kept his vows of peace to the Pilgrims and shared with generosity his food with them, is such a figure. Unfortunately, we have tended to shroud King Philip and Pontiac, as well as other war chiefs, in a mantle of dark conspiracy, when in fact these courageous men were, in their time, leaders of Indian self-determination efforts to rid their land of the white man's culture, which had been so detrimental to their people. Actually, Pontiac was a great leader. His associations with the white people convinced him that the Indian had suffered damage and permanent harm from whites.

Such Indian heroes survive in Indian oral tradition. Oral history tapes, available by the hundreds at several centers for Indian history in the West, show clearly that Indian historical traditions greatly vary from white accounts, although chronology and many points of fact closely tally with white records. What Indians tend to say is that their leaders and heroes were battling all the things Indians complain about today—white encroachment on sacred places; white atrocities in murder, theft, and robbery of personal possessions, cattle, and land; white poisoning of Indian people as well as domestic animals; and white abuses of the soil.

The whole story of the Indian-white confrontation in American history is tragic and full of violence and atrocities. The fact that the Indian and the white man are differing about approaches to our ecological problems is to be expected, for the Indian and the white man have for centuries been disagreeing about religion, government, and a host of other things.[3] So they will continue to disagree, especially about history and the Indian's place in history. Yet at some future time it may be that even Old Posey, an intrepid, determined Ute and the leader of a local war in Utah, as well as Tecumseh, Pontiac, Blackhawk, Sitting Bull, and Chief Joseph, will be recognized as truly national figures in our history. As much as the heroic figure of the scout, the plainsman, the mountain man, and the cowboy himself, these courageous Indians deserve our respect and admiration. For they and all their Indian brethren may have made contributions to our life-style that may ultimately bring about a better America. As Vine Deloria says, the Indian's message is: "Man must learn to live with other forms of life and not destroy it."

NOTES

1. Woodrow Borah and Shelburne Cook argue that the Spanish "disruption of native society and the introduction of new, unusual, and harsh systems of exploitation" also had "a severe effect" on native depopulation in Española. Cook and Borah, *Essays in Population History: Mexico and the Caribbean,* I (Berkeley, 1971), 407–9.

2. Robert Trennert, "The Far Western Indian Frontier and the Beginning of the Reservation System, 1846–1851," Ph.D. dissertation, University of California, Santa Barbara, 1969.

3. Additional bibliographical note: Henry F. Dobyns's "Estimating Aboriginal American Population, An Appraisal of Techniques with a New Hemisphere Estimate," *Current Anthropology: A World Journal of Sciences and Man*, 7 (October 1966), 395–499, contains an addendum with critiques by Bruce Trigger, Harold Driver,

Shelburne F. Cook, and other experts, plus a reply by Dobyns. Harold Driver in *Indians of North America*, 2nd ed. (Chicago, 1969), 63–65, cuts Dobyns's estimates by some 50 percent. Dobyns's theories on Indian depopulation resulting from epidemic diseases are also found in his "An Outline of Andean Epidemic History to 1720," *Bulletin of the History of Medicine,* XXXVII (November–December 1963), 493–515. Shelburne Cook's and Woodrow Borah's brilliant essays on Hispanic population history have appeared in a number of books and articles. Their latest work, *Essays in Population History: Mexico and the Caribbean*, (Berkley, 1971) has a perceptive essay on methodology (p. 73–118) and a painstakingly thorough analysis of data (p. 376–410) behind their estimate of eight million population for Española. The population of Indians in the United States, according to the 1970 census count, was 792,730, excluding Alaskan Eskimos. This figure, of course, does not include hundreds of thousands of Indian people who are blending into the mainstream of American society, many of whom still consider themselves native American Indians but are not counted as such. See U.S. Bureau of the Census, *Census of Population 1970: General Population Characteristics* (Washington, D.C., 1972), 293. One estimate is that there are some five million people of Indian descent in the United States. See William Meyer, *Native Americans, The New Indian Resistance* (New York, 1971), 82.

Duke project tapes at the Western History Center, University of Utah, resulting from hundreds of oral history intervies with Indians of many tribes in variious stations in life, some of them college students and others elderly Indians on reservations (whose replies must be translated into English), record Indian resentment about white encroachment on sacred places, white destruction of the land, white cruelty and duplicity in trade, treaties and war. Many of the accounts reach back into the seventeenth century dealiong with subjects as early as white contacts with Indians, Spanish slavery of Indians, Acoma Indians, and outrages committed by priests. See, for example, tapes numbered 36, 65, 73, 102, 148, 198. One of the most interesting tapes told by a white pioneer on last skirmishes with the Indians in Utah is that of Lynn Lyman, Blanding, Utah, interviewed by Charles Peterson, December 9, 1967, as recorded on Duke Tape No. 668, Western History Utah. Another interesting tape is Duke Tape No.550, an interview with Jim Mike, on Old Posey's death, allegedly by poisoning.

Themes in indian oral history tapes reappear in N. Scott Momaday's *House Made of Dawn*, which tells the authentic and powerful story of modern Indians, who often must live in two worlds, one of which becomes a cycle of dissipation and violence. Momaday's searching essay "The Morality of Indian Hating" in *Ramparts*, III (Summer 1964), 29–40, echoes much of the pride and disillusionment of Indian oral history, as does Vine Deloria's *Custer Died for Your Sins* (New York, 1969).

The impact of the Judeo-Christian ethic upon the Indian and the wilderness is analysed by Vine Deloria in a perceptive essay, "An Indian's Plea to Churches," *Los Angeles Times*, February 6, 1972. Similar concepts are in Lynn T. White's "The Historical Roots of Ecologic Crisis," *Science*, 155 (March 10, 1967), 1203–6. The white man's ruination of the American Southwest, resulting in the black blizzards of the plains, is dramatically portrayed in Pare Lorentz's great film of 1936, "The Plow that Broke the Plains," now available for purchase from the National Archives at an inexpensive price. According to Vernon Carstensen of the University of Washington, "The Plow," once considered an extremely controversial film, was withdrawn by the government from circulation. The case in which it was carried in the 1960s, after it was finally released for showing, bore this warning: "To Be Used for Study as an Art Form Only. Because Subject Matter Is Long Obsolete It Is Not to Be Used for Public Showing." Woody Guthrie's dust-bowl ballads, analyzed by John Opie if Duquesne University at the 1972 Washington meeting of the Organization of American Historians, combine to make a searching social-ecological commentary on the plight of dust-bowl refugees.

Much of the history of land dissipation is recorded in carl O. Sauer, *Land and Life*, (Berkley, 1965), 148 ff. Sauer's *Seventeenth-Century North America, The Land and the People as Seen by the Europeans* (Berkley, 1971) details the ecological impact of European exploration frontiers. Donella H. Meadows et al., *The Limits of Growth, A Report of the Club of Rome's Project on the Predicament of Mankind* (New York, 1972), has a unique assessment of the ecological stage of global equilibrium on pages 156–184. This volume and the rigorous review it received in the *New York Times Book Review*, April 2nd, 1972, by Peter Passell, Marc Roberts, and Leonard Ross, are eye-opening documentaries for anyone interested in controversies on resource exhaustion and pollution crises. The Indian's resistance to the white man's historic conquestof the land is discussed in W. R. Jacobs, *Dispossessing the American Indian: Indians and Whites on the Colonial Frontier* (New York, 1972), 19–30, 126 ff. Modern Indian criticism of the way in which native Americans have been portrayed in secondary textbooks is succinclty set forth by the staff members ofthe American Indian Historical Society in Rupert Costo and Jeannette Henry, eds., *Textbooks and the American Indian* (San Fransisco, 1970). Pages 33, 38, 51, 85, 153, 173, and 216 have critical appraisals of textbooks written by leading American historians. An important new book which provides a sweeping view of the Indian's losing battle with the white man through the courts is Wilcomb E. Washburn's *Red Man's Land—White Man's Law: A Study of the Past and Present Status of the American Indian* (New York, 1971). William Brandon's learned and fascinating volume, *The American Heritage Book of Indians* (New York, 1961), available in an inexpensive paperback edition (1969), is the best overall synthesis of the Indian in our past.

PART III

GREAT INTERPRETERS OF THE FRONTIER EXPERIENCE

CHAPTER SIX

FRANCIS PARKMAN'S ORATION "ROMANCE IN AMERICA"*

Although Jacobs was critical of one-sided heroic frontier history, he still held some early practitioners of American history in high esteem. One of these is Francis Parkman, the patrician romantic who wrote The Oregon Trail *and other nineteenth-century American classics. When Jacobs stumbled on Parkman's oration, "Romance in America," in the Harvard College Library, he found a window into a deeper understanding of the New Englander's writing. The oration demonstrates that, early in his career, Parkman developed romantic appreciation for the North American wilderness. He believed that the key to writing a unique American history lay in setting its past in the vast North American forests, prairies, plains, deserts, and mountains. In his histories, Parkman was interested equally in conquering the "undeveloped" wilderness tracts and in absorbing and enjoying the colors, textures, smells, and sounds of the forests. "Romance in America," reproduced at the end of Jacobs's essay, illuminates Parkman's youthful environmentalism.*

On August 28, 1844,[1] the twenty-one-year-old Francis Parkman delivered a commencement oration on the occasion of his graduation from

*American Historical Review *68 (April 1963), 692–697.*

Harvard. The original manuscript, lodged among his papers in the Harvard College Library and unknown to present-day historians, is in Parkman's handwriting, signed "F. Parkman, August '44."

"Romance in America," as Parkman called the oration, reveals for us the springs from which his later work flowed, providing us with new insight into the romantic concept of history held by one of our greatest historians. Indeed, Parkman's later multivolumed masterpiece *France and England in North America* is, in many respects, a projection of the ideas that so fascinated Parkman in these early undergraduate days. Certainly Parkman's interests crystallized remarkably early in his career, so that the reading of his college days was of real value to him in his literary work. As he himself noted, in a letter written many years later, a literary career early suggested itself as combining his two boyhood loves: love of the forest and of books.[2]

What is known of Parkman's academic career supports his own contention that he was a lover of books. For his scholastic record at Harvard, which was excellent though not brilliant,[3] masks the fact that the prescribed curriculum was only a minor part of his program of study.[4] From his early reading lists and correspondence it is apparent that he carried on a secondary program of reading in literature, ethnology, and history, with particular emphasis on the romantic themes of Francois René de Chateaubriand, Jules Michelet, Sir Walter Scott, and James Fenimore Cooper, all of which readied him for his own literary work.[5]

But though his readings were valuable in preparing him for the oration and his later writings, a six months' grand tour of Europe, taken because of an apparent breakdown in his health, supplied him with the immediate framework for his oration, a comparison of nature in Europe and nature in America. In the early summer of 1844, after returning from Europe, before writing his oration and before his graduation from Harvard in August 1844 (a graduation which seems to have been delayed from the early summer because of his temporary absence from college), he gave evidence of his state of mind with these random jottings in his journal:

> The traveller in Europe.
> Art, nature history combine.
> In America Art has done her best to destroy nature,
> association nothing.[6]

Francis Parkman, ca. 1850

This portrait was made about the time he completed The Conspiracy of Pontiac, *his first major work.*

(Photograph courtesy Wilbur R. Jacobs.)

The quiet beauty of the English countryside had impressed the youthful tourist; in Scotland he was captivated by the "heathery" hills closely associated with the life and writings of Scott. Here were art, legend, nature, and history. America, by contrast, had failed to appreciate the romance of its wilderness heritage, and it was this failure that Parkman hoped to rectify by writing a good—perhaps a great—book on a North American theme, a book that would be recognized as a product of the New World.

It was the sheer grandeur of nature in the wilderness that provided the most impelling motive for Parkman's literary activity. His feeling for this theme is exhibited by the comments of Vassall Morton, the hero of his novel. "Here in America," declares Vassall, "we ought to make the most of this feeling for nature; for we have very little else . . . savageness and solitude have a character of their own; and so has the polished landscape with associations of art, poetry, legend and history."[7] The polished landscape of Europe held little enchantment for Parkman. Rather he turned to the mountains and the virgin forests for his New World symphony. Parkman's fascination with the history of the American forest, stimulated by Cooper's "Leatherstocking Tales," is described in one of his autobiographical letters. In recalling his youth, Parkman (writing in the third person) said: "his thoughts were always in the forests, whose features not unmixed with softer images, possessed his waking and sleeping dreams, filling him with vague cravings impossible to satisfy."[8]

Parkman's treatment of the Indians, unlike his treatment of nature, was far from romantic. He rejected completely the idea of the noble savage, depicting the Indian warrior as a barbarian of the Stone Age. Yet, in spite of this unromantic attitude, his enthusiasm for the Indian was a lifelong affair. Friends of Parkman's college days said that he exhibited symptoms of "Injuns' on the brain" and entertained them with wild tales of Indian scalps, birchbark canoes, and wampum, tales that surpassed anything in Cooper's stories, though they were clearly influenced by them.[9] His *Oregon Trail* of 1849 was a personal narrative of his youthful expedition on the Great Plains and his life with the Sioux, as well as a record of preparation for the historical books that followed, especially the *Conspiracy of Pontiac*. Parkman eventually decided that the story of the French and English in North America would provide the same opportunities for an exciting theme and would be of more interest to general readers than his originally planned history of the North American Indians, with the Six Nations of the Iroquois as a focal point.[10]

Parkman's early fascination with Indian history blended with the interest he developed in studying the heroic figures of Canadian history. The age of the black-robed missionary, the adventurous *gentilhomme* of the forest, and the chivalrous Field Marshal Montcalm held peculiar attraction for him. Painted with splendor on the natural canvas of the primeval wilderness was the history of the soldier in plumed helmet, Indian warriors with barbarous trophies, and the great Jesuit martyrs.

Parkman imagined the past as a kind of theater. The forest was his stage, and historic figures like Robert de La Salle, the Comte de Frontenac, James Wolfe, and Montcalm were "actors" in his drama.[11] Take from his pages in the *History* the backdrop of the woods with its hum of insects, its smell of pine, and its roar of cataracts, and the image dims. Remove the fighting, sweating, and shouting actors, and the interest vanishes, for Parkman did not see his works as a dry chronicle of events, but as drama dependent on people and places. Parkman aspired to create a romantic but authentic image of another age with characters that had the passions of living men and women. His self-assumed task, a half-century of gathering manuscripts, touring historic sites, and writing, was the culmination of a college dream to capture the romance of America's past and make it come to life for others.

That Parkman as a graduating college senior had learned his art is evident from his oration. His prose is made vivid by its appeal to the senses of sight, hearing, and smell. At Lake George a "gentle girl . . . gazes . . . down the Lake"; the sound of a "gun reverberates down the long vista of mountains, and the sullen murmurings dwell for many moments on the ear"; and a raven appears which "once gorged on the dead." The language of the oration reflects also Parkman's youthful affection for poetry, particularly his admiration for the Byronic hero and the forest hymns of William Cullen Bryant. The excellent vocabulary, the graceful sentences merging into smoothly molded paragraphs and transitions, the skillful characterization of the slatternly log-hut pioneer occupied with reading newspapers or cultivating potatoes suggest literary sophistication. But perhaps Parkman is most eloquent when his emotions are aroused, as in praising the unspoiled woods, a subject that was always close to his heart.[12] Yet his style does not violate the beauty of simplicity; it lacks the pedantries and pompous language that punctuated much nineteenth-century prose.

That Parkman gave serious thought to the American Revolution as a historic theme is apparent in his college oration, but he turned away from his subject because it contained, as he wrote, "no display of chivalry or of headlong passion, but a deliberate effort in favor of an abstract principle." What romance was there in a war resulting from a meeting that calmly deliberated and then "voted resistance"?

For him the epochs of the Anglo-French struggle in North America were recalled, as he says in his oration, by the "wild scenery" of Lake George. This glistening jewel in the wilderness, with its rugged shoreline escarpments, pine and hemlock forests, was the "holy lake" christened Lac St. Sacrement by Father Isaac Jogues before he was tormented by the Mohawks. It was the scene of Sir William Johnson's bloody victory over the French and of the massacre of captives following Montcalm's capture of Fort William Henry. Afterward it was the silent witness of the escapades of one of Parkman's favorite heroes, Major Robert Rogers, the colonial ranger.[13] In many respects Lake George was a focus of history for Parkman, a kind of geographical center for his narrative of *France and England in North America*. Even forty years after his college oration Parkman included descriptions of this lake in his *Montcalm and Wolfe.*[14]

Parkman's *History* was recognized in his lifetime as it is today as a gem of historical writing. His books have long been held as models for style and honest research by those historians who aspire to write readable narrative history that has appeal for both the general reader and the specialist. Parkman's college oration reveals in early form the romantic image of history that inspired his later and well-known volumes.

"Romance in America"

> The tourist in Europe finds the scenes of Nature polished by the hand of art, and invested with a thousand associations by the fancies and the deeds of ages. The American traveller is less fortunate. Art has not been idle here for the last two centuries, but she has done her best to ruin, not to adorn, the face of Nature. She has torn down the forests, and blasted the mountains into fragments; dammed up the streams, and drained the lakes, and threatens to leave the whole continent bare and raw. Perhaps the time will come when she will plant

gardens and rear palaces, but the tendency of her present efforts forbids us to be too sanguine.

When Columbus first saw land, American was the sublimest object in the world. Here was the domain of Nature. For ages her forest-trees had risen, flourished, and fallen. In the autumn, the vast continent glowed at once with red, and yellow, and green; and when winter came, the ice of her waters creaked and groaned to the solitude; and in the spring her savage streams burst their fetters, and swept down the refuse of the wilderness. It was half the world a theatre for the operations of Nature! But the charm is broken now. The stern and solemn poetry that breathed from her endless wilderness is gone; and the dullest plainest prose has fixed its home in America.

Once Spain, Italy, England was also a wilderness. The haunts of Nature were *there* in like manner invaded, and her charm broken; but since that remote day, the deeds of many a generation of wise and gallant men have flung around that land the halo of romance and poetry. Its streams and mountains are hallowed by associations that ours have not, and may never have; and the hand of art has polished the rough features of Nature. The warfare of fierce and brave men, seen through the obscure veil of centuries, has given a charm to the Cheviot Hills that will never belong to ours, though our forests have seen struggles more savage and bloody.

The fanciful child, as he journeys through the passes of our northern mountains, looks with awe into the black depths of the woods, and listening to the plunge of the hidden torrent, he recalls the stories of his nursery of Indian wars and massacres. A fearful romance invests all around him, for he associates it with those scenes of horror. And surely the early days of no nation could afford truer elements of romance. They need but to be magnified by superstition and obscured by time. But we live in an enlightened age. History has recorded the minutest circumstance of our fathers' wars; and when we look at the actors, we find the same cool-blooded, reasoning, unyielding men who dwell among us this day—the very antipodes of the hero of romance.

The traveller may pause over the battle-fields of Saratoga or Bennington, and moralize, if he pleases, or give vent to his patriotic ardor. But they have none of the romantic charm, so hateful to the Peace-society. He will not feel the inspiration of Flodden or Otterburn.

Here, on these American fields, was no display of chivalry or of headlong passion, but a deliberate effort in favor of an abstract principle. Cool reason, not passion, or the love, of war, sent the American to the battlefield. When Napoleon placed his brother on the throne of Spain, the Spanish peasant sprang to the gun and the dagger and leaped on the invader with the blind fury of a tiger. The men of New England heard that they were taxed, called a meeting, and voted resistance. Philanthropists may rejoice over the calm deliberation of such proceedings, but the poet has deep reason to lament.

The soldier of the Revolution has handed down to his grandchildren his own cool reasoning temper, so that the traveller finds even fewer elements of the picturesque in the character of the men, than in the aspect of the country. But, perhaps, being young and inexperienced, and having heard that wild men still linger in the recesses of the Scottish Highlands and the mountains of Wales, he imagines that the depths of the yet unwasted forest may contain some form of human nature more strange and wonderful to his American eye. So, with infinite toil, he penetrates to a narrow gap in the woods on the outskirts of civilization;—a small square space hewn out of the forest, and full of the black and smoking carcasses of the murdered trees, while the still living forest palisades the place around. Here dwells the pioneer, in his log-hut. The disappointed traveller finds him like other people, with no trace of primitive ignorance or romantic barbarism. He reads the newspapers; supports Polk and Dallas with fiery zeal; knows the latest improvements in agriculture, and keeps a watchful eye on all that is going on in the great world. Though quite confident in his power to match the whole earth in combat, he has no warlike ardor, preferring to watch his saw-mill and hoe his potatoes, since these seem to him the more rational and profitable occupations. In short, the enthusiast can make nothing of him, and abandoning the thought of finding anything romantic on his native continent, he sighs for Italy, where there are castles and convents, stupid priests, and lazy monks, and dresses of red and green; where people are stabbed with stilettos, instead of being slashed with bowie-knives, and all is picturesquely languid, and romantically useless.

Yet beauties enough be on the northern traveller's path; beauties scarce surpassed on earth, and one spot, at least, whose wild scenery has gained a deeper interest from the early history of his country. A lake which Romish priests, charmed by its matchless beauty, consecrated to the Prince of Peace when that country was an untrodden wilderness, yet which has seen a thousand death-struggles, and been dyed with the blood of legions. To the eye, Lake George seems the home of tranquility and mild repose. The gentle girl sits on the green mound of the ruined fort, and gazes in quiet happiness down the Lake. All is calm and peaceful, yet lovely and wild, by the red light of evening; waters as deep and pure as the eyes of the gazer; mountains whose sternness is softened into a wild beauty. The evening gun reverberates down the long vista of mountains, and the sullen murmurings dwell for many moments on the ear.—There was a time other sounds awoke those echoes,—the batteries of Montcalm; the yells of a savage multitude, and the screams of a butchered army. Blood has been poured out like water over that soil! By day and by night, in summer and in winter, hosts of men have struggled and died upon it. It is sown thick with bullets and human bones, the relics of many a battle and slaughter. The raven that plucks the farmer's corn once gorged on the dead of France, of England, and a score of forgotten savage tribes.

But the Holy Lake is alone. There are other scenes of grandeur and beauty, yet none where associations throng so thick and fast; and as they seem doomed to rest undistinguished in song, we must hope for them the colder honors of prose, and look forward [to] the day when the arts of peace shall have made them illustrious.

F. Parkman, August '44

NOTES

1. The folder containing the MS gives the date of the oration. (Harvard Archives, HUC 6843.55.)

2. Parkman to Pierre Margry, Dec. 6, 1878, *Letters of Francis Parkman*, ed. Wilbur R. Jacobs (2 vols., Norman, Okla., 1960), II, 124.

3. Term Books, Harvard Archives.

4. *Letters of Francis Parkman*, ed. Jacobs, I, xxxiv ff.

5. Parkman's letter of April 29 [1842] to Jared Sparks, written during his sophomore year in college, indicates that he was busily engaged in reading on military campaigns at Lake George during the "Old French War." In his auto-

biographical letter of 1886 to Martin Brimmer, Parkman wrote: "Before the end of the sophomore year my various schemes had crystallized into a plan of writing the story of what was known as the 'Old French War'. . . . " (*Ibid.* 9, 184 n.) Additional evidence of Parkman's youthful interest in early French American history is in the Library Charge Lists, Harvard Archives, and in the notes accompanying Mason Wade's excellent edition of *The Journals of Francis Parkman* (2 vols., New York, 1947).

6. *Journals of Francis Parkman*, ed. Wade, I, 277.

7. Francis Parkman, *Vassall Morton, A Novel* (Boston, 1856), 112.

8. An autobiographical letter to George E. Ellis [1864], *Letters of Francis Parkman,* ed. Jacobs, I, 177.

9. Charles Haight Farnham *A Life of Francis Parkman* (Boston, 1904), 78, in Frontenac ed. of *Francis Parkman's Works* (16 vols., Boston, 1899–1907).

10. Parkman to Abbé Henri-Raymond Casgrain, Oct. 23, 1887, *Letters of Francis Parkman,* ed. Jacobs, II, 213.

11. In his correspondence published in the *Letters of Francis Parkman,* Parkman uses the word "*actors*" in referring to the chief characters of his narrative. The word "*drama*," or the phrase "*dramatic interest*," also appears in the letters in which he discusses plans for the organization his works.

12. "For my part, I would gladly destroy all his works [the *nouveau riche*] and restore Lake George to its native savagery—which shows plainly that you are a better American than I am." (Parkman to Casgrain, Oct. 5, 1892, *Letters of Francis Parkman,* ed. Jacobs, II, 265.)

13. One of Parkman's treasured possessions was an engraving of Rogers (reproduced in *The Conspiracy of Pontiac and the Indian War after the Conquest of Canada* [Boston, 1906], Frontenac ed., I, 266), which he hung in his study at 50 Chestnut Street, Boston. The study was recently dismantled and reconstructed in a similar room at the headquarters of the Colonial Society of Massachusetts, 87 Mount Vernon Street, Boston.

14. See, e.g., *Montcalm and Wolfe* (Boston, 1907), Frontenac ed., II, 181.

CHAPTER SEVEN

FRANCIS PARKMAN—NATURALIST-ENVIRONMENTAL SAVANT*

Francis Parkman was a serious historian-scientist of the natural world. The well-traveled scholar always tried to study firsthand North American geography, flora, and fauna described in his historical narratives. In "Francis Parkman—Naturalist-Environmental Savant," Wilbur R. Jacobs argues that Parkman was a nineteenth-century environmentalist. Although the New Englander hunted wild game, he advocated wilderness conservation and applied ecological concepts to explain how forest preservation benefitted the individual, society, and nature. Jacobs's engaging article explores environmentalism in Parkman's life, writing, and horticultural science.

I am indebted to Martin Ridge, Karen Langlois, Michael Mullin, and Guy Bishop for reading earlier drafts of this paper, which was presented at a session on environmental history at the annual meeting of the Pacific Coast Branch of American Historical Association in August 1991.

Although he wrote in the middle and late nineteenth century, Francis Parkman is still regarded as one of the leading historians of America. Indeed, he is recognized as one of the most sophisticated and able narrators despite the vigorous criticisms of his elitism and social ideas.[1] It is for this reason that his ideas about what we now call environmentalism are of interest. Parkman, of course, could never be accused of being an environmentalist as we now understand the term. Yet there are observable themes of environmentalism in his varied writings, in his experiments with plants, and in his lifelong fascination with natural history. As a leading horticulturalist of his day who, for a time, set up a nursery business, importing and selling plants throughout New England, Parkman was a practical man who fused his talents as a naturalist-horticulturalist with his literary work.[2]

Even though his historical writings have been roundly criticized, he seems to have escaped being attacked for being a naturalist-environmentalist. This omission no doubt stems from the fact that the term *environmentalism* is a twentieth-century expression, but so are such terms of abuse as *racism* and *Anglo-Saxonism*, both of which are often directed at Parkman.[3] Parkman, to be sure, has had a legion of admirers, especially the late Samuel Eliot Morison and Allan Nevins. I knew them both and enjoyed talking with them about Parkman the man and the writer.[4] They liked to comment on Parkman's uncanny ability to provide images of living nature on the printed page. And Parkman himself, if one examines his correspondence and the plethora of his autobiographical essays, his journals, and his statements in the press, saw himself as an honest-broker historian who wrote truthful history about exciting woodland events. Parkman attributed at least part of his success to his studies of "real life" in telling the "history of the American forest." What he wanted to do, he wrote, was to "tell things as they really happened."[5]

Parkman might have added that he designed dramatic narratives, with eyewitness accounts of heroic pioneers battling wilderness and Indians. His novelistic technique, using wilderness as a background for events, has rightly been designated as Parkmanesque. The wildlands could be foreboding, inspiring, magnificent, or serve as a metaphor for humankind. There were humans, like struggling saplings, who were all but suffocated by powerful natural forces. Wilderness tested the fortitude of pioneers. The weak perished and the strong grew stronger. It is safe to say that few, if any, American historians relied so heavily upon environmental themes in their writings.[6]

Parkman's letters to the press during the Civil War era disclose that he viewed himself as a member of a wise patrician class that was needed to provide leadership for his generation. Despite his illness, which he called the "enemy," a neurotic complexity of physical and mental problems including partial blindness, he was well enough to turn out dozens of nonhistorical publications, among them a book on growing roses. He lectured his fellow Americans on a variety of topics, including protection of America's wilderness forestlands, growing lilies and other flowers, and how to write truthful history.

What follows is the argument that the environmental issue can be identified with Parkman in three ways: first, through his activism in writing about wilderness forestlands; second, through his intense identification with wilderness concepts in his multi-volumed *History* and his *Oregon Trail*; and third, through his researches in plant genetics and preoccupation with horticulture.

Let us turn, then, to the first point of argument, that Parkman was a nineteenth-century wilderness advocate. Although his critics continue to hammer on the anvil of his negative views, the fact is that Parkman was extremely positive on certain issues, such as shielding forest wilderness areas from destruction. Although he was a conservative Boston Brahmin, his stance on environmental matters was consistent with the fact that he supported other progressive issues such as civil-service reform and better treatment of Indians as advocated by Carl Schurtz. An aristocratic mugwump in his politics, he was one of those who spoke out in the 1880s about American woodlands being vandalized. On the issue of the need for government regulations one can argue that he was precursor of Charles Francis Adams, brother of his friend, Henry, who led the fight for regulation of railroads.[7] Parkman, though not an activist in opposing expanding tracklines, had a special dislike, even loathing, of the railroads. Not only did locomotive "sparks" cause massive forest fires, their tracks also ruined the beauty of the forest. "I always hated a railroad as the chief instrument for spoiling the woods," he wrote.[8]

Parkman had visceral opposition to civilization's invasion of wilderness. Were he alive we can be sure he would be no friend of the Curry

Company commercialism at Yosemite. In his own time he lashed out at real estate promoters who cut up shorelines of historic Lake George for summer cottages. Complaining to his friend, Abbé Henri-Raymond Casgrain, he wrote, "When I first knew Lake George, the islands of the Narrows were thickly covered by pine, spruce, and fir-trees—as they were represented in the engraving after [William Henry] Bartlett, who saw them about 50 years ago. The *nouveau riche*, who is one of the pests of this country, has now gotten possession of the lake and its islands. For my part, I would gladly destroy all his works and restore Lake George to its native savagery. . ."[9]

Parkman made that comment a year before his death, but in 1885, seven years earlier he made an eloquent assessment of the issue of conservation of America's forests. Indeed, it is significant that Parkman, along with John Muir and George Perkins Marsh, had the foresight in the nineteenth century to write about forest protection for the benefit of future generations. Like them (disappointingly, Parkman's notes or correspondence have no mention of Marsh or Muir) he had a passionate dislike for lumbermen who laid waste to the woods.

But let Parkman speak for himself. His forum was the readership of the *Atlantic Monthly* in 1885. In a review of Charles Sprague Sargent's book on *The Forests and the Census*, he literally exploded into angry but eloquent condemnation of exploiters of America's wilderness.[10] He blasted the ignorance and greed of those who profited from America's "enormous national wealth, which man did nothing to create, but . . . is doing his best to destroy."

There was, he charged, escalating despoliation in the vast American forests, from the hardwoods of the Alleghenies, to the rich pine forests of the South, the redwoods of California and elsewhere. The plundering was the result, he argued, of a wrongheaded attitude toward the wilderness. Although the early settlers "regarded the forest as an enemy to be overcome by any means," their descendants had not yet begun to comprehend that "the old enemy had become the indispensable friend and ally." It was in the "magnificent forests of the Pacific Slope, among the matchless woods of Oregon and Washington," where "the old instinct," Parkman wrote, "springs up again with redoubled force." That instinct was "a selfish love of gain" that also motivated attacks on the "great redwood forests of the coast" and threatened "to deprive posterity of an inestimable possession."[11]

The Mature Francis Parkman
(From a sketch by his son-in-law; photograph courtesy Wilbur R. Jacobs.)

What is fascinating about this plea for conservation is that Parkman explains, with the sure hand of an experienced naturalist, unfortunate causative ecological transformations. As an example, he painstakingly traced cycles of destruction in mountain forests. These particular forests, he explained, are vulnerable because of their fragile toehold on rocky soil. A threat to their very existence came from fire and unbridled cutting. "The forests that cover the tops and sides of the mountains," he said, "generally draw their sustenance from a thin soil formed chiefly of vegetable mould, resulting from many centuries of decay, first of mosses, then of plants and low shrubs, and lastly of trees, each generation contributing something to the support of the next, till the barren ridge where once nothing but a lichen could cling, is able at length to nourish an oak. But, when the forest thus slowly and painfully prepared is swept away by fire the mould burns out like peat, and work of a thousand years is undone in an hour." Fortunately, he added, "in deep soils, on level ground, the mischief is much less."[12]

It was fire, the ax, and still another factor, "the destructive agency of browsing animals," that brought about the disaster and then combined to prevent the forest from renewing itself despite its "unbounded fecundity." In one of the very early protests against the problem of "browsing animals," Parkman echoed John Muir who spoke of browsing sheep as "hoofed locusts." Both Parkman and Muir argued that only the "inaccessibility" of certain mountain forests gave temporary protection. "It is evident," Parkman wrote, "that nothing but the intervention of the state and federal governments can arrest the waste of forests."[13]

"Intervention" to save the forests would also shield much of America from another catastrophic disaster: land erosion. Forest destruction, Parkman maintained, also "means the ruin of great rivers for navigation and irrigation, the destruction of cities located along their river banks." This is what had happened, he said, in northern Italy along the river Po and in the "denudation of the mountainous country" in Spain.

In western America, the Great Divide, which gave birth to the Missouri, the Yellowstone, the Columbia, the Colorado, and the Platte River's North Fork, there were vulnerable forests that covered more than one state. Thus, Parkman said, state governments were often "powerless" to prevent despoliation. The remedy, he said, was federal regulation.

We can summarize Parkman's views as they appeared in a second article, "The Forests of the White Mountains," in the February 29, 1888, issue of *Garden Forest*.[14] The White Mountains, he maintained, were particularly valuable because they provided magnificent scenery and a resulting tourist income. If they "are stripped bare" like some parts of the Pyrenees, a disaster would result in the form of floods, causing "alternate droughts and freshets" from streams flowing into the Merrimac River. Wilderness, he argued, was no longer an enemy but an indispensable ally for the prosperity of the American nation.

While Parkman's two articles on forest conservation probably had little influence on policy makers, we know that the historian-statesman-conservationist Theodore Roosevelt dedicated books to Parkman and shared his enthusiasm for a love of the wilderness life. But Roosevelt, who seemingly devoured Parkman's *Oregon Trail* and his historical volumes, probably never saw Parkman's essays on forest protection; nor is it plausible that John Muir (who had a profound influence on Roosevelt) saw them either.[15] Nonetheless, in the middle 1880s Parkman was among the few Americans who told their fellow citizens that the public domain of wilderness forests was running out. Like George Perkins Marsh and later John Muir, Parkman was an international traveler. And, like them he had firsthand observations of what was happening in foreign countries such as Spain, France, and Italy. He saw the danger in America and advocated federal regulation. As a preservationist he wished to save old-growth mountain forests, and as a conservationist wanted other forest lands to be wisely managed. We should stress that Parkman was not talking about America's bucolic landscapes where the railroad invaded the garden. Parkman's concern was with wilderness. He was one of the first to state that wilderness should not be feared or destroyed. It was part of America's new world paradise and had tremendous value in itself for the individual and for society as a whole.

Where Parkman undoubtedly influenced Roosevelt was on the issue of the inherent value of wilderness environments as ecological conditioners for strengthening those who yearned for the wild spirit in the strenuous life. There are many expressions of this interpretation in Parkman's *History*. Perhaps the best is *The Old Régime* (1874) wherein he extolled the power of the vast interior wilderness in molding the character of Canada's pioneers. "Against absolute authority," Parkman wrote, "there was a counter influence, rudely and wildly antagonistic. Canada was at the very portal of the great interior wilderness. The St. Lawrence and the Lakes were the highway to that domain of savage freedom; and thither the disinfranchised, half-starved seigneur, the discouraged *habitant* who could find no market for his produce naturally enough betook themselves. Their lesson in savagery was well learned, and for many a year a boundless license and a stiff-handed authority battled for control of Canada."[16]

Not unexpectedly this posture of ecological conditioning is found everywhere in the first two volumes of Roosevelt's *Winning of the West*, and there are papers in Frederick Jackson Turner's research notes to show that Turner reviewed Roosevelt's writings on the West and wrote a pilot essay based upon Roosevelt's concept of the heroic hunter type made strong by rustic frontier life.[17] Could there be a historical thread of wilderness character building and ecological conditioning woven in the histories of Parkman that extended itself into the writings of Roosevelt and Turner? And did this cord of fibrous material, twisted into a pattern, project itself into the fabric of western history as written by Frederick Merk and Ray Allen Billington? Perhaps, although that is a different subject.

Parkman, like Roosevelt, tied the strenuous life to hunting and to the historic "hunter type" of early frontiers. And Parkman, like Roosevelt, had an obsession with shooting almost every form of wildlife that came within range of his beloved rifle, "Satan." Close reading of the *Oregon Trail* reveals Parkman's rapturous emotion in shooting and watching the agonizing deaths of bison bulls at the water hole. This preoccupation with killing presents a conundrum in appraising Parkman's role as a wilderness advocate. On the one hand, he wished to preserve and protect the wildland, and on the other, he demonstrated an intense desire to shoot almost everything in it from chickadees to buffalo. He would protect and preserve trees but not animals, probably assuming they would continue to reproduce and survive as separate species. Parkman, unlike Roosevelt the hunter,[18] never commented on this point and never made a statement of regret regarding his slaughter of buffalo on the Oregon Trail. Nor did he write about the possibility that excessive hunting and persistent fur trading for certain species of animals might upset the ecological balance. Yet his writings show that he was keenly aware of the roles of prey and predator, Indians and their hunting-and-gathering patterns, Canadian waste in the reported burning of thousands of beaver skins in price wars, and patterns of forest ecology. He saw animal life on the Oregon Trail as part of an eternal conflict in nature. "From minnows to men," he wrote, "life is incessant war." We see here the view of the despoiler-hunter-predator, tending to ignore ecological balances. He seemed to believe humankind had a kind of entitlement to wilderness, giving them a right to enjoy the bounties of hunting.[19]

One could argue that Parkman was an archetype of the hunter-type of his time, so exemplified by Roosevelt. But unlike Parkman and Roosevelt, John Muir, their contemporary wilderness advocate, saw animals as part of the ecology of the wilderness forests. Muir loathed hunters and particularly trophy killings. His sentiments are set forth in a number of his books, particularly in his *The Story of My Boyhood and Youth*.[20]

There are, aside from the act of hunting, other ways in which Parkman can be identified with the environmental theme, especially in the context of the Leatherstocking wilderness originally popularized by James F. Cooper, one of Parkman's literary models. Parkman's pages are full of vivid, atavistic accounts of what it was like to live out a life in the shaggy forestlands generations ago. Parkman saw himself as a kind of hero-explorer who enjoyed riding over two thousand miles on horseback during his Oregon Trail journey. "The experience of one season on the prairies will teach a man more than a half a dozen in the settlements," he wrote. "There is no place on earth where he is thrown more completely on his own resources."[21] In an autobiographical recollection buried in the second volume of the *Conspiracy of Pontiac*, Parkman rhapsodized, "To him who has once tasted the reckless independence, the haughty self-reliance, the sense of irresponsible freedom, which the forest life engenders, civilization seems flat and stale. Its pleasures are insipid, its pursuits wearisome, its conventionalities, duties, and mutual dependence alike are disgusting. . . . The wilderness, rough, harsh, and inexorable, has charms more potent in their seductive influence than all the lures of luxury and sloth. . . . [H]e on whom it has cast its magic finds no heart to dissolve the spell. . . . [O]ne may fearlessly drink, gaining, with every draught, new vigor and heightened zest. . . . "[22] In these passages we can see the passionate exuberance of the twenty-six–year-old author, emotionally awash in events he chronicled around the story of Chief Pontiac.

One reason the Pontiac narrative has such a sense of immediacy is that the spirit of wilderness magic is everywhere in the narrative. Even white women who had been prisoners of Indians left their village homes after being rescued to return to the wigwam of their warrior husbands. They had no desire, as Parkman wrote, "to abandon the wild license of the forest for the irksome restraints of society."[23]

Parkman's *History*, as Samuel E. Morison has noted, possesses the quality of eternal youth. "That is it," wrote Morison, "the gift of vitality. Parkman's work is forever young, with the immortal youth of art; his men and women are alive; they feel, think and act within the framework of a living nature. . . . Like that sylvan historian of Keats's *Ode on a Grecian Urn*, he caught the spirit of an age and fixed it for all time, forever panting and forever young."[24]

There are, then, explanations for Parkman's triumphs as a writer. One is, without doubt, his extraordinary gift for preserving graceful word-picture images of the formidable wilderness environment that almost overwhelmed pioneers in the seventeenth and eighteenth centuries. Not only was Parkman a lover of what he called "living nature," he also trained himself to be a skilled naturalist, a specialist in horticulture, so he could describe the grandiose wilderness panorama in detail.

Parkman's personal identification with the forests began in his teens. He remembered that at the age of sixteen "a new passion" seized him. Writing in the third person, he declared that "he became enamored to the woods, a fancy which soon gained full control over literary pursuits to which he was also addicted. . . . [H]is thoughts were always on the forests, whose features . . . possessed his waking and sleeping dreams."[25] Those sleeping dreams, as Sigmund Freud has written, had the potential for governing daytime behavior.[26] We are not surprised, therefore, to find the youthful Parkman in 1841 ranging over wild areas of New England and exhausting companions because of his urge, as he said in his journal, "to have a taste of the half-savage life necessary to be led, and to see the wilderness where it was as yet uninvaded by the hand of man."[27] Parkman's wilderness theme fascinated him day and night. Even though his 1846 sojourn on the Oregon Trail nearly cost him his life and his eyesight, he still "looked back regretfully" on leaving "the wilderness behind."[28]

Parkman, as is evident, was not concerned with a pastoral paradise where the machine invaded the garden.[29] To recreate authentic wildland panoramas for his narratives, Parkman visited and revisited actual wilderness sites of main events. If civilization had altered the sites, he did additional work, consulting sources like William Bartram's travel journals. On occasion, he wrote to scientists like Asa Gray. In addition, as his narratives show, original journals of explorers provided a data base in natural history with descriptions of flora, fauna, and geographical landmarks. Besides, he pored over thousands of pages of manuscript documents and made an exhaustive search through contemporary newspapers, pamphlets, and books that could, he said, "throw light on the subject."[30] The youthful Parkman, in short, knew more about Indians, the forest-wilderness, and the history of the early frontier conflict than any other writer of his generation. His historical writings were to have their setting in an ecological wonderland, the original forest of North America.

As Parkman matured he came to know more about "nature" and natural history until he reached the point in early midlife when he was recognized, as we have noted, as both a historian and horticulturalist. His fascination with plant genetics, especially his experiments with species of flowers, roses, and lilies, provides particular insight into his elitist views on the significance of the patrician classes and uses of wilderness. Parkman's roses were true aristocrats.[31] Tamara P. Thornton, in her volume on the elitism of the *Cultivating Gentlemen* of Boston, creates a convincing portrait when she paints Parkman as a man with anxiety over the virility of his patrician class. He used horticulture, Thornton argues, as a means to show that selective breeding creates a new type of aristocracy in plants.[32] Parkman cultivated Brahmin varieties of roses and lilies. In his *Book of Roses* and in his presidential address to the Massachusetts Historical Society, Parkman maintained that the cultivator should strive for "eminence" and not "feeble mediocrity." Horticulture was "no leveler," and families of roses should be recognized as "patricians of the floral commonwealth, gifted at once with fame, beauty, and rank." Horticulture for Parkman, as well as others of his class, showed patterns of heredity that confirmed a kind of aristocratic ethos.[33] There is good reason to believe that Parkman, like his friend Oliver Wendell Holmes, saw the Brahmin class emerging from biological refinement. "Money," wrote Holmes, "kept for two or three generations transforms a race . . . in manners and hereditary culture . . . in blood and bone."[34]

For his plant experiments and his business, Parkman invested thousands of dollars in extensive gardens, a greenhouse, equipment, labor, a gardener's cottage, and plants. In 1862 he decided to enter the nursery business on his three-acre estate at Jamaica Plain with a commercial nurseryman, William H. Spooner. Parkman turned a hobby, which he first viewed as therapy for his illness, into a profit-making operation. He actually made more money in his nursery business than in the sale of his books. Additionally, he won many "premiums," that is, cash-prize awards for his plants, and his *Lilium parkmanni*, resulting from the hybridization of a new variety of lily, was sold for a good profit to an English nurseryman.[35] There can be little question that Parkman's precise skills in floriculture enabled him to include plants and trees in the landscapes that he later described in his historical writings. When he went on field trips to examine the Florida wilderness in 1879 and in 1885 for revisions of his *Pioneers*, he knew exactly what to look for—the "soft, pea green, young foliage of deciduous cypress," and such other trees as maple, ash, gum laurel, myrtle, yucca, mistletoe, wire grass, prickly pear, palmetto, and occasionally "strange climbing plants" that he found difficult to identify. But his journals are filled with dozens of other plants together with a virtual tropical jungle of snakes, alligators, hawks, herons, cranes, ducks, blackbirds, leaping fish, yellow water lilies, fallen logs, black mud, water weeds, and pieces of "shell limestone."[36]

While it is true that the seventeenth- and eighteenth-century environmental scenes Parkman described were often panoramas of what might have been because of changes wrought by nature and by invading white frontiers, he still was able to give his readers an approximation of what the early pioneers faced. His skills as a horticulturalist also enabled him to write with authority and precision on such topics as forest ecology and food chains, although he did not use such terms. Illustrative of his approach was a vivid overview of the New Hampshire forest in a thirty-five–line single sentence in the *Old Régime*, which includes a graphic description of how an ancient "forest devoured its own dead."[37]

Parkman's learning processes as a historian of the early American forest wildlands were part of a lifelong dedicated effort to write a "history of the American forest." His journals, along with some of his early short stories, published in the *Knickerbocker Magazine*, show how the youthful

Parkman transformed his more harrowing American forest experiences into narrative plots. For instance, he could have easily killed himself in rock climbing a peak near New Hampshire's Crawford's Notch. On another occasion he recorded his daring ascent of a steep, crumbling ravine where an avalanche had smothered a pioneer family called the Willeys. That story was retold in a tale called "The Scalp Hunter" and repeated in his novel, *Vassal Morton*. Still later, we find an echo of the episode when General James Wolfe staggered up a steep, watery ravine leading to the Plains of Abraham. Parkman knew about slippery ravines with falling boulders that a heroic rock climber would have to conquer. He had been there before.

With imaginative skill, he designed historical panoramas so vivid that the reader has a sensation of actually moving through the woods along with recognizable historical individuals. We seemingly journey with Parkman through thickets and tangled woodlands, and over lakes, streams, and mountains. We are in backlands of New England with Sir William Johnson's army, in the vast Illinois country with hardy explorer Robert Lasalle, or in the semitropical rain forests of Florida with brave Huguenot pioneers.

We can expect that such an historian would want to write about the need to guard America's precious historic forest lands from vandals. As a self-taught naturalist he recognized immediately the unmatched national treasure in a relic stand of old-growth trees. They were an accessory to human life and a background to human events. The environmental setting in his books was part of the main narrative. He was primarily concerned with a reconstruction of time, place, and conditions in which his characters lived and acted. Nevertheless, as has been demonstrated, wilderness, for Parkman, had particular values, not only for the individual, for society, and for itself. On a metaphorical level, Parkman took strength from the internal stimulation that wilderness gave him and passed on his patrician values to future generations in the same way significant writers have always provided grist for their readers. There are, then, contradictions and complexities in Parkman's environmentalism. He loved the forests and sought to protect them. At the same time he saw the Indian as a savage predator, and he derived personal pleasure in taking on the role of the hunter. Perhaps his most significant legacy was his portraits of what a savage and beautiful place wilderness was before the destruction of the great eastern forestlands.[38] To create his portraits he successfully exploited environmental themes as literary tools in his vivid narratives.

NOTES

1. Wilbur R. Jacobs, *Francis Parkman, Historian as Hero: The Formative Years* (Austin, 1991), 121–148, includes an analysis of Parkman's social perspectives.

2. See note 31 for a discussion of sources on Parkman's significance as a horticulturalist.

3. For a discussion of criticism of Parkman, see the bibliographical note in Jacobs, *Francis Parkman,* 221–228.

4. I had a number of conferences with Morison at Harvard when I was editing Parkman's letters in the late 1950s. In the 1960s at the Huntington Library, Nevins was one of my companions on afternoon walks in the library's gardens and wild areas, still a haven for occasional foxes, coyotes, rabbits, deer, and a plethora of squirrels. Morison was defensive about Parkman's elitism, love of hunting, and views of Indians. Nevins had unbounded admiration for Parkman as a narrator and was responsible for the creation of the Parkman Prize in history.

5. Parkman to Martin Brimmer, 1886; Parkman to Francis H. Underwood, Sept. 7, 1887; Parkman to Abbé Henri-Raymond Casgrain, April 23, 1889, in W. R. Jacobs, ed., *Letters of Francis Parkman* (2 vols., Norman, 1960), I, 184, II, 209, 232.

6. Parkman perfected his craft in writing about environmental themes by testing himself in rigorous expeditions in New England and on the Oregon Trail. An early biographer and personal friend, Charles H. Farnham, pointed out that Parkman trained himself, with the aid of a circus manager, to jump on and off a horse at full gallop almost as if he was preparing for gladiator combat. Farnham, *A Life of Francis Parkman* (Boston, 1904), 130. In autobiographical fictional accounts in *Knickerbocker Magazine* in the 1840s (see, for example, "The Scalp Hunter, a Semi-Historical Sketch," XXV [April 1842], 297–303; "A Fragment of Family History," XXV [June 1845], 504–518), Parkman emerges a hero figure in dangerous rock-climbing episodes and later as a tough, courageous hunter-horseman on the Oregon Trail. These experiences are projected into the bold accomplishments of his heroes such as Robert LaSalle, Robert Rogers, and James Wolfe.

7. On Charles Francis Adams and railroad regulation, see Thomas K. McCraw, *Prophets of Regulation: Charles Francis Adams, Louis D. Brandeis, James M. Landis, Alfred E. Kahn* (Cambridge, Mass., 1984), 1–79.

8. Parkman to Charles H. Farnham, July 25, 1896, in Jacobs, ed., *Letters of Francis Parkman*, II, 195.

9. Parkman to Abbé Henri-Raymond Casgrain, Oct. 5, 1892, *ibid.*, II, 265.

10. *Atlantic Monthly,* LV (June 1885), 835–839. Parkman was not the first person to write on protection for the eastern forests. In the 1860s George Perkins Marsh wrote on ecological catastrophes resulting from forest denudation. In Marsh's *Man and Nature* (1864; Cambridge, Mass., 1965), 219–226, there is a discussion of erosion and deforestation in the Po River area that is more detailed than Parkman's. John Muir's early articles appeared in the *San Francisco Bulletin,* Aug. 13 and 17, 1875; in *Scribners,* XVII (Jan.–Dec. 1879), 55, 260, 411, 478, 644; and in *Century*, XLII (Nov. 1891), 77–97. Another early and eloquent protest against the destruction of American forests was that of Lydia H. Sigourney in her 1845 volume of poetry and prose, *Scenes in My Native Land* (Boston, 1845), 117–120. In her poem "Fallen Forests," she said, "Man's war on trees is terrible . . . [;] he toucheth flame to them and they lie, A blackened wreck." Great forests and gigantic trees were "sacrificed," she wrote. "The intelligent man . . . might surely spare a few . . . a patrimony to an unborn race." William Henry Murray, writing books in the late 1860s on the Adirondack wilderness, and Robert Harris, publishing about this time in the *Overland Monthly*, III (1869), 217 ff, also lamented the decimation of forests. I am indebted to Janice Simon of the University of Georgia for calling my attention to Lydia Sigourney and other early writers on forests.

11. *Atlantic Monthly,* LV (June 1885), 835–839.

12. *Ibid.*

13. *Ibid.*

14. Published in New York.

15. Roosevelt and his protege, Gifford Pinchot, have appropriately been described as "apostles of efficiency." Roosevelt is perhaps best characterized as a conservationist who opposed "land skinning." "I am with," Roosevelt said, "and only with the man who develops the country. I am against the land skinner every time." Quoted in Bret Wallach, *Americans and Conservation: At Odds with Progress* (Tucson, 1991), 50.

16. *The Old Régime* (2 vols., Boston, 1899), II, 198.

17. See Frederick Jackson Turner's 1890 essay on "The Hunter Type," based

almost entirely upon the romantic concept of the "backwoodsman" in the first two volumes of Theodore Roosevelt's *Winning of the West* (4 vols., New York, 1889–1896). Turner's personal copies of volumes three and four, Huntington Library rare book numbers 139455 and 42029, are filled with Turner's marginalia and newspaper clippings tipped into and at the ends of the volumes. Turner's 1890 essay is reproduced in W. R. Jacobs, ed., *Frederick Jackson Turner's Legacy* (Lincoln, Neb., 1977), 153–155.

One of my advanced students recently completed a seminar paper on Theodore Roosevelt as hunter. She stunned our seminar group when she argued that T.R. had a kind of "orgasm" when he shot trophy animals and watched them die. In the ensuing discussion there was general agreement that this viewpoint was correct after hearing Roosevelt's autobiographical accounts of paroxysmal, emotional excitement at the moment of killing. This view of the obsessive hunter type may well apply to Parkman who compared killing bison to shooting enemies on a battlefield. For an analysis of Parkman's preoccupation with martial values, his mental illness, and his role as a voice for the aristocratic masculine tradition, see index entries in Jacobs, *Francis Parkman, Historian as Hero.*

18. Roosevelt, in his *Outdoor Pastimes of An American Hunter* (New York, 1905), 252, opposed the idea that game "belongs to the people," and advocated controlled hunting, presumably for qualified people like himself. For a photo of T.R. standing in front of a pyramid of some dozens of horns and antlers, see "Trophies of a Successful Hunt," opposite p. 277.

19. L. Hugh Moore, in his "Francis Parkman on the Oregon Trail: A Study in Cultural Prejudice," *Western Historical Literature,* XII (Nov. 1977), 185–187, comments on Parkman's recklessness as a hunter.

20. (Boston, 1911).

21. Parkman to his father, Sept. 26, 1846, in Jacobs, *Letters of Francis Parkman,* I, 48.

22. *The Conspiracy of Pontiac and the Indian War after the Conquest of Canada* (2 vols, Boston, 1899), II, 253–256. William H. Murray, contemporary of Parkman, is representative of a wilderness romantic of the nineteenth century. Murray tells a story about "pungent and healing odors" of wilderness spruce and pines that would "heal the diseased and irritated lungs" of a sick young man. After a period in the woods he returned "bronzed as an Indian and as hardy." So potent were the "health giving qualities of the wilderness" that this example of recovery, Murray said, "may seem exaggerated, but it is not." See

Murray's chapter on "Why I Go to Wilderness" in his *Adventures in the Wilderness or Camp-Life in the Adirondacks* (Boston, 1869), 9–14.

23. Parkman, *Conspiracy of Pontiac,* 253.

24. See Samuel Eliot Morison, ed., *The Parkman Reader* (Boston, 1935), x.

25. Autobiographical letter to George Ellis, 1864, in Jacobs, *Letters of Francis Parkman*, I, 176.

26. James Strachey and Anna Freud, eds., and trans., *The Standard Edition of the Complete Psychological Works of Sigmund Freud*, vol. 5: *The Interpretation of Dreams* (London, 1973), 594. Herein Freud discusses the "transfer" of an unconscious wish and the problems of entering the consciousness.

27. Journal of 1841, Parkman Papers, Massachusetts Historical Society. I have used the original manuscript of the journals which have been edited with minor changes by Mason Wade.

28. Parkman, *The Oregon Trail* (Boston, 1900), 464.

29. Roderick Nash, in his pathbreaking book, *Wilderness and the American Mind* (3d ed., New Haven, 1982), 100n, identifies Parkman as an original thinker on values of wilderness.

30. Parkman, *Conspiracy of Pontiac*, I, x.

31. The best discussion of Parkman's leadership as a horticulturist is in Tamara Plakins Thornton, *Cultivating Gentlemen: The Meaning of Country Life among the Boston Elite, 1785–1860* (New Haven, 1989), 144, 207–210, 208n, 226–227. Despite Parkman's lack of formal scientific training, he had genuine triumphs in experimentation with lilies and was nationally known as an expert on roses. His *Book of Roses,* published in Boston in 1866, is still useful for modern rosarians. Today in San Marino, California, at the Huntington Library botanical gardens, his book is often cited in identifying species of roses. Parkman served as president of the Massachusetts Horticultural Society and received no less than 326 awards, including one bronze and sixteen silver medals for his plants. At Harvard in 1871 he was appointed professor of horticulture at the Bussey Institute where he was assigned the task of lecturing to young women on the cultivation of flowers. In his presidential address of 1875 to the Massachusetts Horticultural Society he described horticulture as "an art based on science, or on several sciences" which enabled one "to read the secrets of nature and aid her beneficent functions . . . [, an] ennobling task." The address is quoted in Farnham, *Parkman*, 31–34. Parkman published some thirty articles on growing plants and flowers in the *American Journal of Horticulture* and in the *Bulletin of the Bussey Institution* in addition to his essays on forest conservation.

For a list of these publications, see Mason Wade, *Francis Parkman, Heroic Historian* (New York, 1942), 456; Wilbur L. Schramm, *Francis Parkman, Representative Selections* (New York, 1938), cxxvii.

32. Thornton, *Cultivating Gentlemen*, 208 ff.

33. *Ibid.*, 210.

34. Quoted in *ibid.*, 210 n.

35. *Ibid.*, 226–227; Jacobs, *Letters of Francis Parkman,* I, liv, 98, 138, 146–148, 183, 188.

36. Parkman's journal, 1879 and 1885, Massachusetts Historical Society.

37. Jacobs, *Francis Parkman,* 103.

38. E. Lucy Braun's *Deciduous Forests of North America* (New York, 1964), an older but excellent work on the northeastern forests, has data to demonstrate that Parkman had a firm understanding of forest ecology. See, for instance, pp. 10–38.

In a paper now being prepared for publication, "William Cronon vs. Francis Parkman," I have attempted to compare Parkman and Cronon on early environmental history. This study emerges from the Huntington Library American history reading group's disputatious discussion of Cronon's stimulating book, *Changes in the Land: Indians, Colonists, and the Ecology of New England* (New York, 1983). My conclusions so far indicated that, valuable as Cronon's book is, we still can learn as much or more about the ecology of New England (as well as other parts of eastern North America) from Parkman as we can from Cronon. Cronon's disciple, Timothy Silver, in *A New Face on the Countryside: Indians, Colonists, and Slaves in the South Atlantic Forests, 1500–1800* (New York, 1990), 31 ff, gives an excellent account of food chains, wildlife and plants, and the human impact on the land. He has profitably used such nonhistorical sources as Raymond F. Dassman's *Wildlife Biology* (New York, 1964) and Victor Shelford's *The Ecology of North America* (Urbana, 1963).

CHAPTER EIGHT

FREDERICK JACKSON TURNER

Turner's "Significance of the Frontier in American History"*

Wilbur R. Jacobs has harbored a lifelong fascination for Frederick Jackson Turner, the principal theorist of the American frontier. Jacobs dislikes the oversimplification of Turner's historical thought by his twentieth-century critics and champions. Jacobs considers not only to what Turner wrote but to how he reached his conclusions. In this essay, Jacobs explores the fundamental concept underlying Turner's historical research—his commitment to the broadest possible conception of the past. Turner was critical both of "static" history and of one-dimensional approaches. In his mind, history was a dynamic and complex "process" of change. Breaking down walls between fields and subfields, Turner wanted to understand the interrelationships between frontier, economic, political, social, cultural, and other processes. Only then could the historian, he believed, begin to see historical truth.

The Frederick Jackson Turner papers reveal the problems a great historian had in actually getting his ideas down on paper. Some writers, exemplified by Parkman and Prescott, were fortunate in the possession of literary talents of a high order; indeed they achieved their renown in

*American West Magazine *1 (Winter 1964), 32-35, 78-79.*

large part through their gift of language and evocation. But Turner, though he is capable of admirable lucidity in his essays, and though many of his letters are sprightly and pleasantly conversational in tone, found writing for publication painfully difficult.

Turner's collected papers, his correspondence, research notes, and unpublished essays, reveal that his approach to history was distinctly modern in that it was nonliterary and sought to borrow from the scientific method what history could use. Turner was preoccupied with the massing of evidence, the sifting of data to reach objective conclusions, and the consideration of theoretical formulations and hypotheses. He was above all interested in the interplay of many varied forces—social, political, cultural, economic, geographical—and absorbed in the task of documenting the evolutionary aspects of this interplay, rather than in depicting the character of a man, the mood of a dramatic historical scene, or the appearance of a place.

Turner's professional writings show the influence of this appreciation of the scientific methods. His essays are clear, unadorned historical interpretations, the product of a logical and questioning mind. The stress on ideas, the highly intellectual quality of Turner's writings, might perplex some readers of his works on the frontier and the West, for these subjects seem to lend themselves to a dramatic and romantic theme. But Turner's approach was not romantic, and he wrote of western development not as a romantic saga of American history but as a complex "process" that needed to be understood in all its details.

Turner was more interested in the veracity of thought than in any "artistic" verities or literary "unities." A theme that appears in a number of Turner's letters is that he wished "to be understood" and that he had "no desire," he said, "to prove . . . consistency." In a long autobiographical letter to Merle Curti in 1928 Turner wrote:

> as you know, the "West," with which I dealt, was a *process* rather than a fixed geographical region: it began with the Atlantic Coast; and it emphasized the way in which the East colonized the West, and how the "West," as it stood at any given period affected the development and ideas of the older areas to the East. In short, the "frontier" was taken as the "thin red line" that recorded the dynamic element in American history up to recent times.

Turner's "West," then, was by his own explanation, not the golden one, but as near as he was capable of making it, the real and changing West.

A historical "process," involving, as it does, continuous changes in time and a constantly shifting kaleidoscope of interrelated forces, is a matter much too vast and mobile to be easily grasped. It is incomparably more difficult to described and analyze on paper than a static picture, or even a series of static pictures. The process with which Turner dealt, and that he struggled to get down on paper so it could be understood by others, comprised at least four separate but related movements. These included, as he wrote in an unpublished essay, "(1) the spread of settlement steadily westward, and (2) all the economic, social, and political changes involved in the existence of a belt of free land at the edge of settlement; (3) the continual settling of successive belts of land; (4) the evolution of these successive areas of settlement through various stages of backwoods life, ranching, pioneer farming, scientific farming, and manufacturing life."

The history of the process of westward expansion must, according to Turner, include not only an account of the western occupation of the land and the accompanying evolution of society from pioneer life to an urban manufacturing society but, as he said, "all the economic, social and political changes" at the edge of settlements as these various "Wests" went through a "process" of social evolution. The most gifted and facile writer would have found it an arduous task to write history of such density and complexity!

Had Turner's work been less original, his obstacles would have been less formidable. But his was an independent and creative effort, and he lacked accumulations of background material and published documents which those who followed him have used with profit. Turner's lifetime accumulation of research notes, housed in dozens of bulging file drawers, was not simply an eccentric luxury or the results of a neurotic habit of "a glutton for data." These files were the tools necessary to build the new history. Although Turner declined the responsibility for the so-called "new history," there is reason to believe that his contribution to it was as great as any historian of his generation.

Convinced that most American historians had given distorted accounts of the times they professed to treat, Turner believed he had no real alternative other than to take on the almost overwhelming job of reinterpreting his country's history. He complained that few American historians had written about the West except as "annexation-of-territory history." When he began his researches, Turner perceived that America "beyond the Alleghenies, was almost virgin soil, and that the *frontier process* began with the coast; and that the South also—indeed all the *sections*—needed re-study objectively." The only earlier historian who successfully related "economic and social data to American history" was, in Turner's judgment, Henry Adams who, Turner said, "does much in the first volume of his history with that sort of thing, though he doesn't do it in my way. The tendency was to deal with such topics in one or two separate chapters and then turn, without knitting the two together, to political and diplomatic history. I have tried to keep the relations steadily in mind," Turner wrote, "but it isn't an easy job, and the effort [Here Turner shows himself conscious of his low literary productivity] is sometimes conducive to unwritten books!"

Turner was, of course, aware that in his attempts to counteract the static histories of the past he was in danger of being misunderstood. "Undue emphasis," he thought, had been placed upon his interest in western history. Critics often failed to grasp, he wrote in a letter of 1925 to Arthur M. Schlesinger, that his theory of the frontier and the "process" of westward expansion was an attempt to show how "the interior" was "a necessary element in an understanding of the America of the time." This "interior" Turner saw as "a modifying element," a "mixing bowl," and as "a parent region for much of the mountain states and the [Pacific] Coast." Turner thought of the new "Wests," in various stages of social and economic evolution, as colonies of the East, and collectively he sometimes referred to them as America's "interior." The actual frontier, however, had to be distinguished from the westward movement as a "process." "Of course [the] Frontier and [the] West are not identical," Turner wrote in one of his letters to Schlesinger, "but I used [the] Frontier as (so to speak) the barometric line that recorded the advance of settlement, the creation of new Wests, not merely as the area of Indian fighting, vigilantes, annexations, etc." Similarly, in writing to Carl Becker he pointed out that his self-assumed task of treating precisely those areas most neglected by historians could be easily misinterpreted:

Frederick Jackson Turner Contemplating the Old and New Western Historians on Wagon Roads West
(Cartoon by Wilbur R. Jacobs.)

> Although my work has laid stress upon two aspects of American history—the frontier and the sections (in the sense of geographical regions, or provinces . . .),—I do not think of myself as primarily either a western historian, or a human geographer. I have stressed these two factors, because it seemed to me that they had been neglected, but fundamentally I have been interested in the inter-relations of economics, politics, sociology, culture in general, with the geographic factors, in explaining the United States of to-day by means of its history thus broadly taken. Perhaps this is one of the many reasons why I have not been more voluminous!

Turner's view of "broadly taken" history was certainly not easy to grasp or to put down on paper. With some justification Turner objected to his conception of the frontier being used "as a man of straw to be pummeled." It is not surprising that he roused himself from a sickbed in December 1931, shortly before his death, to dictate a long letter to a critic who inferred that he was guilty of "spreading an error" on frontier history.

Turner disliked oversimplification involved in being labeled, or "tagged," one thing or another. His concept of history was so broad that he even objected to being classified as "a Western historian." When a distinguished scholar told him that he behaved like a sociologist, not a historian, Turner answered, "I didn't care what I was *called*, so long as I was left to try to ascertain the truth, and the relation of facts to cause and effect in my own way." Turner would naturally enough object to the idea, voiced by Charles A. Beard and Arthur M. Schlesinger, that he stressed an economic interpretation of history. Writing to Schlesinger in 1922, Turner said that he did not know if his frontier studies represented "fundamentally an economic interpretation." "There is in this country," he continued, "such an interrelation of ideals, economic interest, frontier advance (or recession, if you prefer), and regional geography, that it isn't easy to separate them." While Turner believed that his essays on the West showed an "appreciation of the economic changes," they were, he wrote, "more related to the influence of free lands and the frontier in the large sense." "The truth is," he added, "that I found it necessary to hammer pretty hard and pretty steadily on the frontier idea to 'get it in,' as a corrective to the kind of thinking I found some thirty years ago and less." He hoped, he wrote Schlesinger in this letter of

1922, to spend his later years adding a "companion piece (the *Section*) to the Frontier." If he lived, Turner said in this interesting letter, he would "attempt a coordination of these old and new viewpoints in a general sketch of our history, emphasizing the dynamics rather than the statics: the genetic element, and the *flow* of it all."

The task that Turner chose for himself, then, was to have his new viewpoints recognized and to combine them in a dynamic history based upon study of interrelationships and what he called "the genetic element." He was well aware that the "frontier process" he sought to describe was a phase of American history related to other historical processes "essential" to understanding our past. These other processes, such as "the evolution of sectionalism," "the evolution of political institutions," "the evolution of a composite, non-English nationality," "industrial transformations," and "the slavery struggle and the Negro suffrage problem," show Turner's appreciation of aspects of Darwinian theory. Beyond this, not only did Turner see that the frontier process might have applications in countries other than his own, in Russia and in South America for example, but he recognized that the history of all the world might be considered in terms of the "processes" of history. Turner wrote Carl Becker:

> The "frontier" process is one which applies to certain portions of Old World history, as well as to that of the New, and sometime it will be worked out thus. And beyond all this is the conception of history as a complex of all the social sciences. The conception of the One-ness of the thing. As you intimate, this is rather paralyzing to begin with (and I can't claim authorship of it)! But it does help to know that these subjects are tied together and to deal with a phase of the whole, realizing it is only a phase.

It is small wonder, considering the vastness of Turner's chosen task (involving a conception of history that even Carl Becker found "rather paralyzing"), that Turner was sometimes depressed at his failure to produce more in the form of polished printed results. "In truth," Turner wrote, "there is no single key to American history. In history, as in science, we are learning that a complex result is the outcome of the interplay of many forces. Simple explanations fail to meet the case."

The task of writing the new history, taking the broad Turnerean view in interpreting our past, was surely an undertaking of the first magnitude. Turner himself acknowledged that "the man who does general history on these lines must indeed be a genius; but," he added, "with some equipment in the other social fields and some knowledge of the scientific method and tools he should be able to consult the special works intelligently so as to proceed not too narrowly in the orientation of his subject and the development of it."

Turner in giving special attention to one important phase of "general" American history was fascinated by the great mass of human action on the frontier and in the West. He saw American society clothed in its spiritual and material Old World inheritance, a captive of customs, suddenly thrust into an infinitely complex environmental and social setting on the raw frontier. It was here that the "bonds of custom" were broken by the frontier process of social change.

With the greatest earnestness, Turner sought to understand the relationship between the confused memory of the past and the intricate reality of the present. Despite the magnitude of the problem he felt an obligation to leave the results of his investigations in printed form. "We must build foundations," he wrote, "and furnish *real* bricks for those who come after us, and profit by our mistakes and half-sight—and there is a real danger of merely thinking 'by and large.'"

By the time of his retirement from Harvard, when his tenure as a research associate began at the Huntington Library in the Far West, Turner had formed conclusions concerning the results of his theorizing. Writing to Merle Curti he said:

> My work, whether good or bad, can only be correctly judged by noting what American historians and teachers of history (see College & Univ. catalogues of the later eighties) were doing when I began. . . . The attitude toward Western history was at the time largely antiquarian or of the romantic narrative type devoid of the conception of the "West" as a moving process—modifying the East, and involving economic, political and social factors.

In speaking of his methodology Turner added,

> Mine, if I have any, has been largely unconscious . . . the search beyond the skyline for new truth, and the use of such methods of getting there as immediate needs & resources permitted.

While occupied in a lifelong search for this "new truth," Turner seems to have acquired a feeling of guilt at his failure to leave more for later historians. But we, looking back on his life, can see that his publications were held in check by precisely those qualities that made him a historian of such enduring value—his awareness of the enormous scope and complexity of history, which precluded rash and casual judgments, and his strong sense of responsibility to his profession, which caused him to busy himself so magnificently with its future practitioners, his students, represented by Merle Curti and Carl Becker, who became leaders of a great host of distinguished disciples. Turner's main contribution in his theorizing was perhaps best phrased not by a disciple but a critic, Charles A. Beard, who wrote in a letter of 1928, "It was Mr. Turner who lead in putting history on a scientific plane."

ACKNOWLEDGMENT

In the course of a long preoccupation with Turner's writings and his papers at the Huntington Library, at Harvard, at the University of Wisconsin, and at other document repositories I have contracted intellectual debts that are difficult to repay. I am particularly indebted to the brilliant essayists on Turner: Carl Becker, Merle Curti, Fulmer Mood, Avery Craven, Ray Billington, and Norman Harper. Most of all, however, this paper, which is part of a book-length study on Turner's thought and his significance as a teacher and historian, is based on what he said about himself, especially in his later years. This information was culled from a large mass of Turner's unpublished essays and lectures, his research materials, and his correspondence with Carl Becker, Merle Curti, Arthur M. Schlesinger, Arthur H. Buffinton, Max Farrand, Isaiah Bowman, William E. Dodd, Charles A. Beard, Edgar Eugene Robinson, Harry Elmer Barnes, James Harvey Robinson, and others. The large portion of this manuscript material is described in "The Frederick Jackson Turner Papers in the Huntington Library," by Ray A. Billington and Wilbur R. Jacobs, *Arizona and the West*, II (Spring, 1960), pp. 73-77. In his letters Turner sometimes referred his students to his American Historical Association presidential address, "Social Forces in American History" (published in *The Frontier in American History*, New York, 1920, pp. 311-34), for what he called, "my general ideas of the scope and methods."

CHAPTER NINE

FREDERICK JACKSON TURNER—MASTER TEACHER*

Historiographers generally ignore the teaching accomplishments of their historical subjects, but Wilbur R. Jacobs believed that Frederick Jackson Turner's classroom teaching was crucial to understanding his written history. Recognized by his students as a "great teacher," Turner wove together his classroom and research interests, each venue raising questions to be explored in the other. Training his pupils in the "new history," he encouraged them to accumulate large pools of data and think of history in the broadest terms. Not surprisingly, Turner shined brightest in seminars, where he demonstrated his vast knowledge of historical data and sources, trained young historians, and continually refined the historical method. When Jacobs wrote this essay, he believed that Turner applied the method of multiple working hypotheses to historical analysis but later revised this conclusion.

The story of Frederick Jackson Turner as a great teacher is a gradually fading memory, and the fact that Turner made a powerful impact upon the historical world as a teacher is often overlooked by today's historians. Much emphasis has been placed upon a few of his writings, espe-

Reprinted from Pacific Historical Review *23 (February 1954), 49–58.*

cially the essay, "The Significance of the Frontier in American History," but little attention has been devoted to the living teacher and the attitudes he manifested in the classroom and in the seminar. Only by giving proper balance to Turner the teacher and Turner the writer can his correct place in American historiography be ascertained.

The difficulties in securing data regarding Turner as a teacher are manifold for one who has not had the opportunity to take his seminars. Many of Turner's students, who are still active in the historical world, however, have been most cooperative in giving information concerning their former teacher by means of conferences and correspondence. Two of these students have also made their original classroom notes available, and Turner materials in the Huntington Library, the Houghton Library at Harvard, and the Wisconsin State Historical Society have also been utilized.[1]

Out of the testimonials of his students, his correspondence, and his writings emerges the many-sided genius of Turner. His searching, suggestive mind, his magnetic personality, and his warm sincerity made lifelong friends of his students. Not only students, but fellow professors and two future American presidents, Theodore Roosevelt and Woodrow Wilson, were among his devoted disciples. Wilson set high value on Turner's counsel and friendship and wanted to have him as a university colleague. His letters to Turner contain numerous requests for aid in locating sources, maps, and other data. Theodore Roosevelt was never an intimate friend, but he also felt the force of Turner's vast knowledge of American history and declared that Turner was "a master of the subject."[2]

Although Turner's influence as a teacher penetrated far beyond the classroom, it was here that many of his original ideas were first put into words. It is true that practically all of his views as a teacher are found somewhere in his writings; but Turner did not repeat what he had written when he was teaching. Rather he discussed his plans and research problems and gave "fuller development from later studies and writings which he had not yet published."[3] "It is doubtful whether he ever really separated teaching and research."[4] This manner of teaching baffled many who attended Turner's undergraduate classes; but, as one eminent student pointed out, it baffled none who ever took work in his seminars.[5] Turner, for example, incorporated his research into his teaching by frequently emphasizing recurring trends in history, by citing developments within the sections, and by using illustrations for various sectors of the frontier.[6]

Despite this interest in the frontier and in the section, Turner encouraged his students to work in other fields. Indications of his many research interests can be found in the wide variety of subjects that former students investigated. The master took justifiable pride in the diverse and scholarly accomplishments of his students. He was regarded as a friend, teacher, and critic until the very end of his days.[7]

Turner's students maintain emphatically that their master did not have a restricted interpretation of American history. The late Joseph Schafer declared that "the West was merely his specialty as a student of America."[8] Merle E. Curti has pointed out that Turner suggested and explored almost every field of historical research or significant interpretation of history.[9] Indeed, Turner urged his students to study world history because no one nation could be understood without considering its connection with other countries.

With complete lack of dogmatism, Turner viewed history as an inexhaustible subject. He was ready to examine, to adopt, or to reject any historical hypothesis.[10] Although he did not emphasize the concept of the multiple hypotheses in his writings, Turner's students were made aware of this idea. This approach was not original with Turner, and he always acknowledged his debt to the distinguished geologist and president of the University of Wisconsin, Thomas Chrowder Chamberlin, whose study on multiple hypotheses dealing with the origin of the earth, prompted Turner to apply the idea to history.[11] Turner recognized the need for applying every possible hypothesis to a historical problem. Like the eminent geologist, Turner was dealing with causality. Both men urged a wide range of investigational effort in their researches. Turner was concerned with hypotheses to explain the interacting forces in modern society, and Chamberlin was concerned with the origin of the earth and the development of the solar system. Illustrative of Turner's method of investigating a hypothesis was his use of statistics to correlate political and social behavior on physiographic maps.[12] Thus, from data on physiographic maps, he was able to show the correlation between Congressional votes for a tariff on raw wool and the westward migration of the sheep industry.[13] Here were the factual data to support a hypothesis.

The reasons for Turner's original approach to the study of American history have been the subject of much interest to historians.[14] The fact that he had only one-third of a year in the study of American history under Professor William F. Allen at Wisconsin before he went to Johns Hopkins University might have been an advantage. When Turner arrived at Baltimore in 1888, he had not been indoctrinated into any particular American history teacher's mode of thinking. Turner, in short, had to work out his own salvation.[15]

This lack of training undoubtedly aided Turner to develop his breadth of view and unique interpretation of American history. His training under William F. Allen, a Latin scholar and a medievalist, caused the future historian to interpret American history in the spirit of the medieval historians who had to deal with the growth of major institutions in their formative period. As a mature historian, Turner once stated that he believed all the social sciences were one, and related to the physical sciences.[16] This view, he said, considerably influenced his writing and his teaching.[17]

In working with his students, Turner placed foremost importance upon the study of the present. To explain the problem of modern America, he encouraged the investigation of the interplay of forces in American society. Here were the clues to the "life, ideals, and problems of the present." The national spirit of Americans, which Turner called "Uncle Sam's psychology," was a complexity whose roots were formed in the fertile soil of the federation of different sections of the nation.[18] Sectional rivalries for institutional and political control of pioneer areas were, for example, rivalries that supplied the clues to much of our political party history.[19]

Another inclusive topic that helped to explain the present was the modifying influence of new environments. Turner urged his students to examine a wide variety of environmental factors and the effect of these factors upon the growth of American society. Examples of modifying factors that students analyzed are soil, climate, topography, Indians, native plants, and mineral resources.[20]

While Turner pointed out the significance of modifying influences in new environments, he did not neglect to recognize the persistence of habits of emigrants from older regions in unsettled areas. Of course, the two ideas seem contradictory, but it is evident that the former concept was more prominent in his teaching.[21] Here again one finds the theme of multiple hypotheses.

It follows that students were stimulated to approach history in an analytical fashion. In dealing with biographical material, for example, preeminent value was attached to the study of a leader's environment.[22]

The society in which a great leader lived, the lesser men whose support he needed, and the opposition who modified his policy were all conditioning influences affecting the career of a statesman. Turner's concept of a biographical study left out much of the traditional minutiae of personal life. He did stress, however, that students should "dig deeply" and reject subordinate details.

This master teacher had no particular philosophy of history or method; but he did emphasize "constant alertness" and evidence. Any interpretation that Turner encouraged his students to make was based upon the careful evaluation of data. Each problem involved different factors and new evidence. Though he had no formula for sifting facts, Turner often talked of different kinds of evidence, including "the reputation of witnesses for intelligence, veracity, opportunity to know, bias, and self-interest."[23] In this search for truth he was willing to encourage the examination of unconventional source materials in the fields of sociology, economic geography, and demography, the statistical study of populations. The wide range of materials that Turner utilized to collect his data impressed his students. Many of these students learned, like their master, to make generalizations based upon widely scattered facts. In the opinion of Woodrow Wilson, Turner was one of the few men who "could combine the large view with the small one."[24]

Turner was an inspiring teacher because of his intellectual integrity and his knowledge of his subject material. It was in the seminar that Turner was at his best. Here was the real historian and the teacher. Helpful, encouraging, and friendly, he was generous with his time.[25] Turner was not sarcastic in criticism of student work or of other writers. The emphasis was upon constructive work. His enthusiam had a magical quality that gave significance to the history of every township, county, territory or state. The students caught the spirit, and works like the *Documents Relative to the Colonial History of New York* seemed to be "invested with a color and charm."[26]

Turner usually brought a stack of notes to his seminar, and he sometimes had difficulty in sorting his material. These notes were usually from source materials as distinguished from secondary authorities and textbooks. Turner, however, did not scorn the historians who had "built before." He demanded wide reading in the standard works and at the same time encouraged independent study. The seminar was a workshop, and the master and the students worked together.

Professor Frederick Jackson Turner Conducting an American History Seminar in an Alcove of the Wisconsin Historical Society's Library in the State Capital during the 1893–1894 academic year.

Turner had a few women doctoral students, but most women terminated their studies at the master's degree, which qualified them to work in libraries or teach in noncollege posts.

(Photograph courtesy State Historical Society of Wisconsin, WHi[X3]46720.)

Each student chose or was assigned a topic within the general framework of the course. The student later presented his study to the group, as with most seminars, and a discussion followed, supplemented by comments from Turner. "A good paper got attention; a bad one was its own condemnation and merited little comment."[27] One year, for example, the seminar worked on the period 1820–1840, each person studying one state, and each was regarded as an authority on his locale. On other occasions, students investigated topics like agriculture, banking, land policy, or slavery. Each seminar student was expected to hand in a report on some limited topic in his own field. In addition, the members of the group were asked to write individual correlative papers that evaluated the other reports and showed relation to their own work. This was the substance of Turner's seminar procedure, which might be called a "joint venture of students and teacher in a search for truth."[28]

Turner was an exceedingly good listener during student reports. He gave these student efforts his utmost attention and frequently took notes. Each student thus had the feeling of contributing to the group. This feeling of cooperation was enhanced by the fact that Turner brought his own notes and plans to the seminar.[29]

It has been noted that Turner frequently had his seminars work within a narrow chronological span. In the early years while teaching at Wisconsin, he conceived the idea of securing a basic knowledge of American history by studying successive periods beginning with Virginia colonization. Student reports were a part of this system. Turner followed this plan all through his Harvard years, omitting seminar studies in the Civil War period because of the emphasis that era had already received from historians.[30]

Turner's notes were arranged according to this scheme, the material within each drawer in several filing cabinets being arranged chronologically under general geographical headings. For example, drawer one contains numerous notes, masters theses, maps, and monographs dealing with the period from 1492 to 1690 under such topics as Virginia, New England, and the Atlantic Plains. Drawer two covers the period from 1690 to 1763. Homer C. Hockett of Ohio State University, one of Turner's former students, also used this scheme in his notes, which were recently given to the University of California. Thus, Turner influenced his students to use his methods, and his students contributed to the master's knowledge of history by their researches. This system that Turner used explains his almost phenomenal knowledge of the sources of American history. Such a detailed knowledge, covering the whole panorama of American development from the seventeenth to the twentieth centuries, made his lectures particularly meaningful.

On the lecture platform Turner was particularly effective with the mature and thoughtful student. The alert listener at once discerned that here was a "first-class mind" in operation. The lectures were analytical and punctuated with the generous use of maps and slides. The spontaneity of Turner's presentation and his occasional reference to his notes precluded what might be called a "polished lecture."[31]

At Harvard, where Turner's courses were for upper classmen and graduates, he usually entered the classroom informally, with a briefcase stuffed with notes and illustrative material. This latter was chiefly maps and charts, mounted upon linen, and some of these maps were in connected sections to be used before the classes. The lectures were generally well-organized, interpretive, and occasionally salted with "digs, as in the case of New England's neglect to expand her frontier northward."[32]

Students like Herbert Eugene Bolton, Edgar Eugene Robinson, Avery Craven, and many others were impressed with Turner's warm personality.[33] Of medium stature, blond, handsome, and endowed with a melodious voice, which contributed to his outstanding record in oratorical contests in his undergraduate days, Turner was a pleasing and effective lecturer, "always without ostentation and bombast." He was never overbearing or dogmatic, and he had a real sense of humor that he quietly displayed.

"Freddie" was the name that many students affectionately and egotistically called him, but never to his face.[34] They admired and respected him because of his modesty and intellectual honesty. His lectures told the students that the teacher was till trying to learn; this characteristic of the lectures won student confidence, and it encouraged independent thought and inquiry.

Perhaps the best way to illustrate the kind of material Turner presented in his lectures is to quote from a set of notes taken by Homer C. Hockett in the summer of 1902 at the University of Wisconsin. Hockett took the notes in shorthand and then immediately typed them out in sentence form. The following paragraphs from the History of the West course probably contain the exact words used by Turner in a lecture on July 9, 1902:

At the last meeting of the class, we completed our general survey of the settlement of the interior, the Piedmont region and the Valley, and I tried to bring out the fact that a peculiar people had been established in the region of the Valley and the Piedmont; that this was the seed-plot for the pioneer backwoods stock that went on to conquer the wilderness in the period between then and, roughly speaking, 1830, when the New England stock began to share the task with them in large numbers; and this was peculiarly a pugilistic frontier stock. The Scotch Irish have been likened to the finely tempered steel edge of the woodsman's axe. They furnished the cutting edge of the frontier line.

In illustration of the development of this region and its antagonism with the coast, I wish to have you consider the situation politically between the years, roughly speaking, 1765 and the outbreak of the Revolution. I desire, I say, to illustrate the contest, the antagonism, between the tidewater region and the Piedmont region in that period with the view of developing the fact that a new peculiar region had grown up that tended toward democracy and that type of American democracy had a tenacious earnestness about its democracy. We shall find the struggle between the coast and the interior a continuous one. It is going on at the present time. Ask the man of Kansas or Nebraska what he thinks of Wall Street, and you will get a reply couched in language which the man of the Piedmont region might have used with respect to Charleston and the coast in the days that we are speaking of.

The first grievance of the frontier is in connection with the apportionment of the legislature. Here I simply call your attention again to the tables which I threw on the screen at the last meeting, showing that in Virginia and the two Carolinas the coast was greatly in excess of its proper power in the legislature, if we judge proper power to be based upon population rather than on property. These grievances existed especially in Pennsylvania, Virginia, and the two Carolinas.[35]

Turner continues by discussing the problem of representation in detail as it concerned Pennsylvania and the Carolinas. The lecture ends with an analysis of the Regulator revolt, and throughout the lecture Turner refers to documentary sources and monographs.

If Turner were here today, and if he could read this paper with all its praise and idealization, his innate modesty would probably cause him to minimize his accomplishments as a teacher. When he received the copy of Carl Becker's beautifully written biographical essay now found in Odum's *American Masters of Social Science,*[36] Turner complained that his former student saw him in a "golden light."[37] Carl Becker, in justifying a student's adoration of his master, replied: "As to idealization of your personal influence—well, no doubt but this doesn't bother me. . . . Even if we saw you in some golden light, . . . there was something there calculated to make us see you thus magnified. . . . But I know how it strikes you, because many of my students think I have done great things for them, whereas (so it seems to me) [I] have done nothing but sit around and look wise."[38]

Turner undoubtedly deserved the "golden light" recollections of his students. It is a matter of common knowledge that he was selected by the Council of the American Historical Association as one of the two most eminent American historians.[39] Such recognition was of Turner's remarkable accomplishments as a teacher as well as of his skillful and interpretive writings.

NOTES

1. The bulk of Turner's papers at the Huntington Library is not available at the present time. The Houghton Library collection is a small but important selection of correspondence contained in three large folders. The collection at Wisconsin includes several letters relating to Turner as a teacher.

One set of notes was taken by Horace J. Smith of Los Angeles, California, while he was a student in Turner's History of the West course at Harvard in 1911. The other notes were taken by Professor Homer C. Hockett when he was acting as an assistant to Turner in the year 1902. These notes were made at the request of Turner, and a copy was given to him.

2. Theodore Roosevelt to Frederick Jackson Turner, Nov. 4, 1896, Turner Papers, Houghton Library. See also Woodrow Wilson to Frederick Jackson Turner, Dec. 27, 1884, *ibid.*, and Woodrow Wilson to Reuben G. Thwaites, Dec. 26, 1889, *ibid.*, for Wilson's dependence upon Turner.

3. Comment by Thomas P. Martin. Frederick Merk indicated by correspondence that there were no points emphasized in the lectures that were not evident in Turner's writings. It is of interest to note that Turner was concerned about the duplication of his general ideas in his published essays. See his letter to Guy Stanton Ford, Nov. 27, 1920, Turner Papers, Wisconsin State Historical Society.

4. See Edgar Eugene Robinson's statement on Turner in the *North Dakota Historical Quarterly,* IV (1932), 259–261.

5. *Ibid.*

6. This information appears in Thomas P. Martin's comments on Turner and in an essay that Herbert Eugene Bolton sent to me entitled, "Turner as I Remember Him." Professor C. B. Goodykoontz indicated by correspondence that Turner was more systematic in his coverage of the history of the West in his lectures than he was in his writings.

7. In a letter to Edward T. Hartman on Dec. 29, 1925, Turner Papers, Wisconsin State Historical Society, Turner wrote that much of his work was to be found in the achievements of his "seminarians." Also see Max Farrand, "Frederick Jackson Turner at the Huntington Library," *Huntington Library Bulletin,* No. 3 (1933), 163; Joseph Schafer, "The Author of the Frontier Hypotheses," *Wisconsin Magazine of History*, XV (1931), 86–103; Edward Everett Dale, "Memories of Frederick Jackson Turner," *Mississippi Valley Historical Review* XXX (1943), 339–358; Edward Everett Dale, "Turner—the Man, the Teacher," *University of Kansas City Review* (Autumn, 1951), 18–28.

8. Joseph Schafer, "Turner's America," *Wisconsin Magazine of History,* XVII (1934), 447–465.

9. Merle E. Curti, "Frederick Jackson Turner," *Instituto Panamericano de Geografía e Historia, Comision De Historia*, Num. 96 (Mexico, 1949), 10. Hereafter referred to as Curti, "Frederick Jackson Turner."

10. See footnote 8; Joseph Schafer, "Turner's Frontier Philosophy," *Wisconsin Magazine of History*, XVI (1933), 451–469; Max Farrand, "Evaluation of Turner's History of the West Course," an unpublished manuscript in my possesion.

11. Curti, "Frederick Jackson Turner," 14. See also Merle E. Curti, "The Section and the Frontier in American History: The Methodological Concepts of Frederick Jackson Turner," in Stuart A. Rice, ed., *Methods in Social Science: A Case Book Compiled under the Direction of the Committee on Scientific Method in the Social Sciences of the Social Science Research Council* (Chicago, 1931), 357; Homer C. Hockett also emphasized Turner's interest in the theory of multiple hypotheses.

Thomas Chrowder Chamberlin (1843–1928) was a leading authority on glaciers and the geology of Wisconsin; in 1887 he was called to the presidency of the University of Wisconsin, and in 1893 he founded the *Journal of Geology.* Chamberlin was noted for his desire to examine all data. He recognized the need for the application of every possible hypothesis for the study of geological problems like the origin of the earth.

12. Curti, "Frederick Jackson Turner," 37.

13. *Ibid.*

14. See Fulmer Mood's "Turner's Formative Period," in Fulmer Mood and Everett E. Edwards, eds., *The Early Writings of Frederick Jackson Turner* (Madison, 1938), for a complete evaluation of Turner's formative period of intellectual development.

15. According to Guy Stanton Ford, and other early Turner students, the master teacher frequently gave his students a great deal of freedom in their research work. It appears that many of these students also had to work out their own "salvation."

16. This statement was made by Turner to one of his students.

17. *Ibid.*

18. Turner's autobiographical letter to Constance Lindsay Skinner March 15, 1922, *Wisconsin Magazine of History,* XIX (1935), 97. In this letter, Turner points out the close relationship between the study of the American frontier advance and the development of new sections. He talks in terms of North and South, and East and West, and "many inter– and intra-states sections." He also emphasizes that there is much unworked material on the influence of the frontier on foreign relations, and he refers to his articles in the *American Historical Review* on French policy, X, 249, and the *Atlantic*, XCII, 676, 807, which illustrate his interest in this subject.

19. Homer C. Hockett made this comment in his evaluation of Turner as a teacher.

20. *Ibid.*

21. *Ibid.*

22. According to Fulmer Mood, this was Turner's approach to biographical study.

23. Herbert Eugene Bolton, "Turner as I Remember Him."

24. Joseph Schafer, "Turner's America," *Wisconsin Magazine of History*, XVII (1934), 450. For additional material relating to Woodrow Wilson's opinion of Frederick Jackson Turner, see *ibid.*, XXVI (1943), 470–471; Ray S. Baker, *Woodrow Wilson's Life and Letters . . .,* (Garden City, 1927), II, 124–125.

25. See Avery Craven's essay on Frederick Jackson Turner in William H. Hutchinson, ed., *The Marcus W. Jernegan Essays in American Historiography* (Chicago, 1937), 267. It states: "No other historian of the day inspired in others so

much original investigation." Colin B. Goodykoontz, Frederick Merk, Merle E. Curti, Edgar Eugene Robinson, Herbert Eugene Bolton, Thomas P. Martin, Homer C. Hockett, and Edward E. Dale emphasized that Turner was generous with his time. Hockett stated that frequently at the University of Wisconsin when engaged in his researches, Turner would be interrupted by students. Looking through his opaque glass window, Turner could see the student who was interrupting his work and almost invariably he would take time to give consultations. See also Max Farrand, "Frederick Jackson Turner at the Huntington Library," *Huntington Library Bulletin,* No. 3 (1933), 163.

26. See Carl Becker's essay on Frederick Jackson Turner in Howard W. Odum, ed., *American Masters of Social Science, An Approach to the Study of the Social Sciences Through a Neglected Field of Biography* (New York, 1927), 282, hereafter cited as *American Masters of Social Science.* The above essay is also printed with some variation in Carl L. Becker, *Every Man His Own Historian. Essays on History and Politics* (New York, 1935), 191–232.

27. See note 23. For an interesting account of Turner's seminars, see *American Masters of Social Science,* 284–285. Becker relates the following incident concerning a seminar report on the Mexican War: "Here before me, for example, are the notes, easily contained on one sheet of paper 6x8, a two-days' report on the Mexican War. 'Turner asks: "By the way, Mr. Rogers, what exactly are the Biglow Papers?" Rogers says: "The Biglow Papers—(Hesitates, seems a little dazed, has at last a happy inspiration), "Why, the Biglow Papers are—a well known work by—a famous author." Hilarious laughter, led by Turner, who then explains Biglow Papers' Don't myself know B.P. Remember look up and read B.P. Lowell, J.R."

28. Louise Phelps Kellogg emphasizes the comradeship between professor and students in Turner's seminars; see her essay, "The Passing of a Great Teacher—Frederick Jackson Turner," *Historical Outlook*, XXIII (1932), 270–272.

29. *American Masters of Social Science,* pp. 273–318.

30. According to Fulmer Mood, Turner did have one of his seminars work on the Civil War period. See also Max Farrand, "Frederick Jackson Turner at the Huntington Library," *Huntington Library Bulletin* No. 3 (1933), 163. As indicated in Max Farrand's essay, Turner's notes are now deposited with the Huntington Library.

For Turner's evaluation of A. B. Hart's method of note taking, see Joseph

Schafer, "The Author of the Frontier Hypothesis," *Wisconsin Magazine of History,* XV (1931), 99.

31. *Ibid.*, 100.

32. Thomas P. Martin made this statement in his summary on Turner as a teacher. Turner occasionally in his lectures appeared to neglect the presence of his class by giving the impression that he was thinking aloud. In short, some of his lectures were almost like a soliloquy. Professor Homer C. Hockett attested to this in an evaluation of Turner's technique of lecturing.

33. Herbert Eugene Bolton emphasizes this point in his essay, "Turner as I Remember Him."

34. *Ibid.* Carl Becker wrote that Turner was known as "Old Freddie Turner."

35. Quoted from the set of notes taken by Homer C. Hockett in the History of the West course at University of Wisconsin in the summer of 1902. A copy of these notes has been given to the University of California by Professor Hockett.

36. See note 26 for the full title of this work.

37. Turner's letter to his former student is not available; but it is evident that Turner objected to Becker's idealization.

38. Carl Becker to Frederick Jackson Turner, Ithaca, New York, May 18, 1927. This letter was found attached to Turner's copy of *American Masters of Social Science*, which is now in the Huntington Library.

39. Curti, "Frederick Jackson Turner." Francis Parkman was the other historian selected by the Council of the American Historical Association.

CHAPTER TEN

TURNER'S METHODOLOGY

Multiple Working Hypotheses or Ruling Theory?*

At the turn of the century, Turner was publically and professionally recognized as a founder of the "new history," which sought scientific understanding of the historical process. In "Turner's Methodology," Jacobs explores how the master analyzed his historical data. Turner was a disciple of geologist Thomas C. Chamberlin's "Method of Multiple Working Hypotheses." When analyzing data, Chamberlin argued, the scientist must apply multiple working hypotheses to generate lines of inquiry and to guard against the adoption of "pet theories," which easily become "ruling theories." Jacobs demonstrates that, although Turner advocated the application of Chamberlin's methodology, his research and writing were guided by his two major historical ruling theories—frontier and section—throughout his professional career.

Frederick Jackson Turner told his students that his own work had been strongly influenced by the idea of multiple hypotheses in scientific research postulated by the geologist Thomas C. Chamberlin. Homer C. Hockett recalled that Turner often had stressed Chamberlin's concepts

**Reprinted from* The Journal of American History *54 (March 1968), 853–863.*

in lectures and in the seminar at Wisconsin; and Merle Curti, who studied under Turner at Harvard, has also emphasized Turner's reliance on the concept of multiple hypotheses.[1]

The question of the extent to which Turner actually adopted Chamberlin's methods is, indeed, an important one, not only because Turner believed Chamberlin's methodology to be as consequential for the serious study of history as it was for geology but also because Turner's own approach to historical research has exercised so great an influence on modern historical writing.[2] Since Turner hoped that the procedures that he himself developed would help to make historical scholarship a more impersonal field of endeavor and perhaps bring historical scholarship closer to (even within the magic circle of) the sciences, it is relevant to examine the extent to which his own work satisfied the demands that he made upon scientific scholarship, as well as the extent to which it satisfied the criteria set forth by Chamberlin.

In 1882, the year that Chamberlin left Beloit College to become president of the University of Wisconsin, Turner was an undergraduate at Wisconsin, where later he earned his master's degree. After a period at Johns Hopkins, where he secured his doctorate, Turner returned to Wisconsin as an assistant professor in 1889.[3] This was also the year when Chamberlin read a paper at a meeting of the Society of Western Naturalists entitled "The Method of Multiple Working Hypotheses," to which was appended the provocative subtitle, "With this method the dangers of parental affection for a favorite theory can be circumvented."[4] Revising the piece for the *Journal of Geology* some years later,[5] Chamberlin wrote that it had "been freely altered and abbreviated so as to limit it to aspects related to geological study."[6] This latter version has been reprinted a number of times, and in its approach to geological research, it is still valid today. Thus, Marland P. Billings of Harvard in his volume, *Structural Geology*, cites Chamberlin's essay and notes that:

> Above all, the field geologist must use the method of "working multiple hypotheses" to deduce the geological structure. While the field work progresses, he should conceive as many interpretations as are consistent with the known facts. He should then formulate tests for those interpretations, checking them by data already obtained, or checking them in the future by new data. Many of these interpreta-

tions will be abandoned, new ones will develop, and those finally accepted may bear little resemblance to hypotheses considered early in field work.[7]

In 1888, the year before Chamberlin read his perceptive essay, Turner and Chamberlin had collaborated in writing an essay on Wisconsin for the ninth edition of the *Encyclopaedia Britannica*.[8] Four years later, at the meeting of the American Historical Association held at the World's Fair in Chicago, Turner presented his epoch-making paper, "The Significance of the Frontier in American History"; in 1910 he left Wisconsin for Harvard, where he continued to tell his students about his indebtedness to Chamberlin. The geologist, in the meantime, had left Wisconsin for the University of Chicago to assume the chairmanship of its newly established department of geology.[9]

A comparison of the original and the revised versions of Chamberlin's paper on multiple working hypotheses reveals that he remained consistent about methodology. The ruling theorist, Chamberlin insists, was constantly exposed to the temptation of formulating premature conclusions on the basis of facts revealed by his investigations. "The mind," he notes, "lingers with pleasure upon the facts that fall happily into the embrace of the theory, and feels a natural coldness toward those that assume a refractory attitude."[10] But even when the investigator attempts to maintain an impartial attitude, the problem of "unwarranted vacillation" exists—the danger that, in considering the various working hypotheses, the investigator, unwilling to take a stand, will sway from one line of policy to another.[11] This tendency to vacillation and procrastination springs from the difficulties inherent in the search for knowledge—"the imperfections of evidence," the vast unknown in science and in life itself that make the investigator wary of judging too quickly.[12]

Despite the limitations implicit in the method of multiple working hypotheses, Chamberlin held that its use was obligatory in the scientific world; and, indeed, it might even be applied to "the varied affairs of life," where almost unlimited opportunities for its application—especially in teaching—might be found.[13] What scholars needed to learn, according to Chamberlin, was to recognize the proper time for decision making—to withhold judgment until sufficient evidence has accumulated to justify conclusions.[14]

Chamberlin's references were always to "multiple working hypotheses." By stressing the word *working*—which Turner seems never to have used in this context—Chamberlin placed particular emphasis on the tentative nature of his hypothesis; for the scientist, he insists, must maintain complete detachment from "pet" theories, which may all too easily become "ruling" theories. "The working hypothesis," Chamberlin argues, "differs from the ruling theory in that it is used as a means of determining facts rather than as a proposition to be established."[15] The function of the working hypothesis was to suggest lines of inquiry: "The facts are sought for the purpose of ultimate induction and demonstration. . . " Under the ruling theory, by contrast, the facts are sought in order to support the theory.[16]

Time and time again, Chamberlin warns against the tendency of the working hypothesis to slip quietly into the role of the ruling theory. A hypothesis by its very nature may quickly become a "beloved intellectual child" and finally "a controlling idea." The ideal investigator, Chamberlin notes, should be "the parent of a family of hypotheses"; among these are "intellectual children (by birth or adoption)." Some may die before reaching healthy maturity, but all must survive "the results of final investigation." "The effort," Chamberlin concludes, "is to bring up into view every rational explanation of new phenomena, and to develop every tenable hypothesis respecting their cause and history."[17] Chamberlin's method, as his disciple Bailey Willis observes, "calls upon the student to lay aside a natural preference for the theory which seems plausible and to consider as sincerely that which holds out small promise of development."[18] The investigator must at all costs avoid being entrapped by the ruling theory.

Obviously, it was not simply Chamberlin's scientific approach that fascinated Turner; methodology alone did not make him a lifelong disciple of the geologist. In fact, he was very much interested in the influence of geological and geographical factors on the history of a country, and Chamberlin's work provided him with a rich source of relevant geological information. Therefore, it is not surprising that Turner's offprint of Willis's article dealing with Chamberlin's theories concerning the effect of atmospheric carbonic acid on glaciers is heavily underlined and replete with marginalia.[19] This interest in a field not normally associated with American history repeatedly emerges from Turner's work. For example, in a lecture delivered at the California Institute of Technology, entitled "The Sectionalism of Politics," Turner pointed out that:

> in the election of 1856, the counties that voted in favor of Fremont almost exactly coincided with this second glacial ice sheet . . . the land of the basin of the Great Lakes and the prairies rejected by the Southern settler and occupied by Greater New England.[20]

Turner's geological interests were also evident in the works of his students—such as the doctoral dissertation of Orin G. Libby, "The Geographical Distribution of the Vote of the Thirteen States on the Federal Constitution, 1787–8,"[21] which proved to be a milestone in the geological-geographical approach to American history and which developed out of Turner's lectures on the constitutional and political history of the United States.[22] The technique of map analysis used in this study later proved helpful to Turner in his sectional studies.

Many of the concepts that historians have associated with Turner—the fall line, Appalachian barrier, geological-geographical syndrome—were based upon the study of the topographical maps published by the United States Geological Survey. Turner found these maps especially helpful; and he used them for charting election results, population movements, and studying the correlations between types of soil and areas of white illiteracy.[23] He even had an annotated set of topographical maps placed in the library at Harvard, and students of his "History 17" course were required to consult them.[24] The lectures of the geologist Charles Richard Van Hise, his Madison colleague and friend, provided Turner with another useful source of geological inspiration. He took copious notes on subjects that interested him; one of these was Van Hise's lecture on the Gulf Plains, which described not only the rocks and soil of the area, the deposits made by the Mississippi and its tributaries, the crops and natural resources of the region, but also its history.[25] Van Hise touched on the theme of America's vanishing mineral and lumber resources—a topic that became almost an obsession with Turner after his retirement, when he was gathering material for a book on problems resulting from overpopulation, war, and the depletion of food supplies.[26]

Of course geology was only one of the disciplines that, according to Turner, had something to contribute to the study of history. When Turner was told that he behaved like a sociologist, he answered that he did not care what he was called:

> so long as I was left to try to ascertain the truth, and the relation of the facts to cause and effect in my own way. . . . I have been dubbed by the Sociologists an economic determinist (which I am not!), by the geographers as a geographer, etc. . . . and I as you perhaps recall, valued Chamberlin's paper on the Multiple Hypothesis, which I have aimed to apply to history as he to geology.
>
> Perhaps at the bottom the belief that all the social sciences were one, and related to physical science has influenced my work.[27]

The difficulties facing the would-be scientific scholar in his attempts to achieve "historical fairmindedness" were readily apparent to Turner: "unconscious interpretation, selection, emphasis, . . . conscious, but unsuccessful interpretation, selection and emphasis. 'We're all poor critters!'" he once wrote,"—especially F.J.T."[28] The modesty of this statement was perfectly sincere, but it was meant to emphasize the rather obvious fact that truth is hard to apprehend. In fact, Turner firmly believed that he was applying Chamberlin's scientific approach to history. "One must adopt," he noted in a "Lecture on Political Maps," "the geologist's use of the multiple hypothesis to explain complex areas; and must not attempt to give a decisive reason for the political complexion of a given country at a given election."[29] That Turner emphasized the scientific nature of his work was understandable, for he was indeed more objective, more farsighted, more open-minded than those who had dealt with American history before him.

Yet at the same time, Turner cannot be accepted at face value when he insists that he is applying Chamberlin's scientific methodology to history. There is the distinct impression that Turner confused Chamberlin's multiple working hypotheses with multiple causation. Chamberlin felt that one value of the theory of multiple working hypotheses was that it could lead to the discovery that some phenomena are the result of a number of causes (but this did not have to be so), whereas the ruling theory and even a single working hypothesis lead to a mono-causational interpretation. When examining Turner's writings, one notes that though Turner called upon the social sciences in his interpretations of American history and made use of the comparative method in arriving at conclusions, nevertheless, his work as a whole was very much shaped by the two theories for which he is usually remembered—the frontier theory and the theory of sections. Turner used these theories to construct a framework within which the seemingly pointless minutiae of history take on form and become usable ingredients for the historian's analytical alchemy. American history, as viewed by Turner, consisted of a long process of change, an extended adjustment to conditions on the American continent. The various societies created across the country as the settlers moved west, lured by the promise of plentiful free land, passed through the several states separating a primitive society from a highly developed one; and as they did so, they developed sectional interests and sectional characteristics. But the pattern that helped Turner pick his way through the past vitiated his claim to be following the theory of multiple working hypotheses. Brilliant as he was, Turner's work reveals rather too clearly his affinities with those unscientific investigators whom Chamberlin scorned—those who practiced the "ruling theory" approach.

But in all fairness to Turner, to judge the value of his work by the degree to which he followed, or failed to follow, Chamberlin's method is to bind him unnecessarily in a straitjacket—although in one of his own making. In fact, Turner can be admired for a number of different reasons—for his ingenuity and freshness of mind, for his willingness to erase the boundaries separating disciplines, for the wealth of ideas scattered through is writings (ideas taken up and expanded by more recent historians), for the clarity with which he makes us aware of the complexity of his subject matter, for his unerring instinct for what is important, for his skill in emphasizing the larger trends—not, however, for the scientific character and completeness of his work. On the contrary, as Avery Craven points out in his essay, "Frederick Jackson Turner, Historian,"[30] Turner must be excused certain unscientific traits of mind, since these are associated with intellectual characteristics that we value in him:

> There was something of the poet and much of the philosopher about Turner. He had the ability to see deep into the meaning of things and the power to catch the universals. This did not weaken his capacity for scientific research nor lessen his interest in details, but it did cause him to emphasize trends and flavors, to attempt to deal with intan-

Frederick Jackson Turner in His Study at Home in Cambridge, Massachusetts, 1923
Turner retired from Harvard a year later.
(Photograph courtesy Huntington Library, San Marino, California.)

> gibles, to sweep over minor things in the effort to get at the larger truths. This method has its dangers if history is to be viewed as a pure science and not as a mixture of science and art.[31]

Moreover, Turner himself recognized the special, at best semiscientific, nature of history, which requires that each new generation write "*the history of the past anew with reference to the conditions uppermost in its own time.*"[32] An "ultimate" history—a really scientific history—he therefore knew could never be written. Even if the nature of history had not precluded a strictly scientific approach, Turner's enthusiasm for his subject matter and his patriotic pride in his country would have made him the wrong man to attempt the job. "This progress of society from pioneer life on a seashore, to the colonization of successive sections, and to the final occupation of a continental empire," he states, "is one of the most wonderful chapters in human history."[33] Since Turner's feelings lent a somewhat rosey hue to his discussion of the frontier and its inhabitants, it is hardly surprising that his frontier theory became widely known and widely accepted. Here was the new national history, the theoretical basis for understanding the American character. The special qualities of American democracy, it was felt, could be explained by the frontier experience and by the evolution of sections.

Early in his career Turner concluded that if he was to use the vast resources of the disciplines outside history, he had to identify guidelines to cope with the jungle of data that confronted him.[34] While he was in the process of establishing a kind of order, he evolved his theories of the frontier and the section; but at the same time, he quickly recognized the importance of other theories and hypotheses. For example, Turner's commonplace books disclose that he was an enthusiastic reader of the works of Charles Darwin and Herbert Spencer.[35] In his drafts of student orations[36] as well as in his later essays, he placed considerable emphasis on the doctrine of evolution, especially when he stressed that American society went through "successive stages of social evolution."[37] He thought of society as an "organism," and history was actually the self-consciousness of that organism.[38] In addition, Turner at times emphasized the economic theory of history. In his teaching he gave so much importance to economics that he felt obliged to tell one class that "this is not a course on that subject, economic history."[39] Still another idea that Turner utilized in writing and in teaching is now called the culture-concept, or theory, the idea that old and new emigrants brought their particular ethnic heritage into developing areas. In his last book, *The United States, 1830–1850: The Nation and Its Sections,* Turner developed this concept in discussing the growth of regional cultures.[40] These and other theories, then, were incorporated into Turner's brilliant theory of sections, which is, in a sense, a means of objecting to theories, for it posits the complexity of historical causation. It was through this theory that Turner explored the interrelationships of social, geographical, political, evolutionary, cultural, and economic forces in the development of American society.

Thus it was natural for Turner to encourage his students and asssociates to investigate a whole range of historical interpretations, although he did not develop them as his own hypotheses. Turner's letters abound with many friendly suggestions for investigation as, for instance, his encouragement of Aurthur M. Schlesinger in 1922 to give further attention to "the phenomena of great city development and the results and problems in many fields thereto."[41]

Turner, of course, was not responsible for the oversimplified linear interpretation of American history, which a few of his followers have foisted upon him.[42] Yet it is true that he devoted his life to gathering the material with which to support his own interpretation of the American past. His belief that he had carried Chamberlin's method over to historical research was quite sincere; but, in fact, he seems to have confused the interdisciplinary, comparative method of investigation with the methodology of multiple working hypotheses. A further complication is that he seems to have also confused the idea of multiple causation in history with both the comparative method and Chamberlin's technique. The approach that he did pioneer has, nevertheless, widened enormously the scope of historical investigation; the Chamberlin approach, which he failed to apply, may well be inapplicable to history. However, Chamberlin's plea for very high standards of objectivity on the part of researchers will be echoed by most modern historians, as indeed it was by Turner himself.

Turner's research notes, file drawers of lectures, and various unfinished manuscripts (covering the whole period of his productive life, from the 1880s to the 1930s) support the conclusion that he was primarily a goal-oriented type of researcher. In the dozens of file drawers and in the large map collection at the Huntington Library there is no single research project or paper or chapter of a projected book that is completely divorced from the sectional-frontier theme. In one of these file drawers there are some seventy unpublished pieces, all concerned with some aspect or other of sectionalism, many of them cannibalized from his published work.[43] Indeed, almost everything Turner wrote after 1893 was concerned with sectionalism—a concern that developed out of his frontier hypothesis. He even applied some of his ideas about sectionalism and political parties to problems of international organization and suggested the need for international political parties as a step toward eliminating war.[44] He also began a study of the origin of the city, which he traced, in part, to economic and social changes on the frontier and in the section.[45] In later life he placed increasing stress upon comparative and statistical methods of study but his focus remained unchanged.

Turner's lifetime work, then, was built around the dual theme of frontier and section. Certainly this theoretical scaffolding provided Turner with valuable opportuntities for the study of the interplay of historical forces and for the reinterpretation of various phases of history. Nevertheless, Turner's contribution remains but one of many explanations of our national development, a "ruling theory" that has, with justice, been strongly contested by a number of other scholars.[46] Turner seems never to have regarded his hypotheses with the scientific detachment of a Chamberlin. He never really considered his central ideas mere hypotheses to be tested and compared with other theories—and possibly rejected. On the contrary, Turner was a loving father to his theories: to them he devoted his life, and for their sake no scholarly effort in gathering data was too great.

NOTES

1. Homer C. Hockett to author, Oct. 29, 1951. Frederick Jackson Turner to Merle Curti, Aug. 8, 1928, Frederick Jackson Turner Papers (Henry E. Huntington Library and Art Gallery).

2. Emphasis on Turner's "profound" influence and his use of the "multiple hypothesis in history" appears in a eulogy by Samuel Eliot Morison, Aurther M. Schlesinger, and Frederick Merk, "Minute on the Life and Services of Frederick Jackson Turner," July 14, 1932, Turner Papers. Turner's application of the multiple-hypothesis idea and the importance of his association with Thomas C. Chamberlin is noted in two brilliant articles: Merle Curti, "The

Section and the Frontier in American History: The Methodological Concepts of Frederick Jackson Turner," Stuart A. Rice, ed., *Methods in Social Science: A Case Book* (Chicago, 1931), 357; and Merle Curti, "Frederick Jackson Turner, 1861–1932," *Probing Our Past* (New York, 1955), 36, 39–40. Turner's indebtedness to Chamberlin is also briefly discussed in Fulmer Mood, "The Development of Frederick Jackson Turner as a Historical Thinker," Colonial Society of Massachussets *Transactions,* (Boston, 1943), XXXIV, 286, 319; Ray Allen Billington, *America's Frontier Heritage* (New York, 1966), 19; Wilbur R. Jacobs, ed., *Frederick Jackson Turner's Legacy: Unpublished Writings in American History* (San Marino, Calif., 1965), 40–41; Maurice M. Vance, *Charles Richard Van Hise, Scientist Progressive* (Madison, 1960), 61; William Coleman, "Science and Symbol in the Turner Frontier Hypothesis," *American Historical Review,* LXXII (Oct. 1966), 28n–29n.

3. Chamberlin to Turner, Feb. 27, April 10, 1889, Turner Papers.

4. *Science*, XV (Feb. 7, 1890), 92–96; reprinted in *Science*, 148 (May 7, 1965), 754–759.

5. *Journal of Geology*, V (Nov.–Dec. 1897), 837–848.

6. *Ibid.*, 837n.

7. Marland P. Billings, *Structural Geology* (New York, 1954), 3. Commenting on Chamberlin's essay, a noted biophysicist wrote: "it should be required reading for every graduate student and for every professor." John Rader Platt, *Step to Man* (New York, 1966), 28.

8. *Encyclopedia Britannica* (25 vols., Edinburgh, 1888), XXIV, 616–619. The "Geology" section of the article (pp. 616–617) has Chamberlin's initials: "T.C.C."

9. George L. Collie and Hiram D. Densmore, *Thomas C. Chamberlin . . . and Rollin D. Salisbury, A Beloit College Partnership* (Madison, 1932), 1–47.

10. *Journal of Geology*, V (Nov.–Dec., 1897), 840.

11. *Science* (May 7, 1965), 758.

12. *Ibid.*, 759.

13. *Ibid.*, 758.

14. *Ibid.*, 759.

15. *Journal of Geology*, V (Nov.–Dec.) 1897), 842. Slightly modified in *Science* (May 7, 1965), 754.

16. *Journal of Geology*, V (Nov.–Dec. 1897), 842; *Science* (May 7, 1965), 755.

17. *Science* (May 7, 1965), 756.

18. Bailey Willis, "Climate and Carbonic Acid," *Popular Science Monthly* (July 1901), 242–256. Turner's marked copy is Rare Book 246668 (Huntington Library).

19. Rare Book 246668 (Huntington Library).

20. "American Political Sectionalism," Pasadena Lecture, Feb. 28, 1928, Turner Papers.

21. Orin G. Libby, "The Geographical Distribution of the Vote of the Thirteen States on the Federal Constitution, 1787–8," *Bulletin of the University of Wisconsin: Economics, Political Science, and History Series,* I (Madison, 1897), 1–116.

22. See Mood, "The Development of Frederick Jackson Turner as a Historical Thinker," 331.

23. Turner Map Collection, Turner Papers.

24. *Ibid.*

25. "Turner's Notes on the Van Hise lecture, 1898—Nov., Gulf Plains," Turner Papers. Turner's mutual interests and lifelong friendship with Charles Richard Van Hise are discussed in Vance, *Charles Richard Van Hise*, 61–62, 69–70, 72, 74, 118. See also Mrs. Turner's "Journal of a Camping Trip," Aug. 10–Sept. 10, 1908, Turner Papers.

26. Turner's annotated copy of Raymond Pearl's "The Population Problem: and Walter E. Willcox's "On the Future Distribution of White Settlement" (both in the Huntington Library reference copy of *Geographical Review*, XII [Oct. 1922], 636–45, 646–47) and his notes evidence his concern with this problem.

27. Turner to Curti, Aug. 8, 1928, Turner Papers.

28. Turner to Curti, Aug. 27, 1928, *ibid.*

29. "Notes for a Talk on Political Maps, circa 1920," *ibid.* See also Jacobs, *Frederick Jackson Turner's Legacy,* 70–73.

30. O. Lawrence Burnette, Jr., comp., *Wisconsin Witness to Frederick Jackson Turner: A Collection of Essays on the Historian and the Thesis* (Madison, 1961), 100–116.

31. *Ibid.*

32. Everett E. Edward, comp., *The Early Writings of Frederick Jackson Turner, With a List of All His Works* (Madison, 1938), 52. The italics are Turner's.

33. Frederick Jackson Turner, "The Development of American Society," *Alumni Quarterly of the University of Illinois,* II (July 1908), 120–136.

34. Turner's 3 x 5 reference cards, some of which were compiled in the early 1890s, show his interdisciplinary approach to research. Even the early cards, dog-eared and written in lavender ink, are chronologically arranged under topics on the growth and expansion of American civilization in the eighteenth and nineteenth centuries.

35. Commonplace Book No. 2, Vol. III, Turner Papers.

36. In his draft of an oration entitled, " 'Imaginativeness of Present' or its general worth as contrasted with past," Turner wrote: "New poets will read a lesson from Spencer & Darwin & sing Man and Nature. Evolution . . . is now in the intellect. . . ." *Ibid.*

37. Jacobs, *Frederick Jackson Turner's Legacy,* 166.

38. *Ibid.*; Curti, *Probing Our Past,* 36.

39. Jacobs, *Frederick Jackson Turner's Legacy,* 82. In a brief but provocative essay, "Class and Sectional Struggles," Turner explored economic and social conflicts that could result from a Negro revolution in the South and the triumph of "Bolsehvistic labor ideas." *Ibid.*, 77–78.

40. It was published after Turner's death by Max Farrand, Avery Craven, and Turner's secretary, Merrill H. Crissey. This book, which Turner had under preparation for some twenty years but left unfinished, analyzes each of the larger sections separately with special attention to cultural developments. For example, see Turner's discussion of culture in "The South Atlantic States," *The United States, 1830–1850: The Nation and Its Sections* (New York, 1938), 200–209. Curti's astute appraisal of the book is in *Probing Our Past*, 51. For a discussion of the culture-concept, or theory, and Turner's contribution toward its development, see Caroline F. Ware, ed., *The Cultural Approach to History* (New York, 1940), 10–14, 235–236, 279, 304–306.

41. Turner to Schlesinger, April 18, 1922, Turner Papers.

42. Still appearing are books and articles that ostensibly suggest disagreement with Turner but are actually based upon an analysis and restatement of his ideas. For example, the basic Turnerian themes of frontier mobility and abudant free land underlie David M. Potter's *People of Plenty, Economic Abundance and the American Character* (Chicago, 1954); Walter Prescott Webb's *The Great Frontier* (Boston, 1952); and George W. Pierson's articles, "The M-Factor in American History," *American Quarterly*, XIV (Summer 1962), 275–289, and "A Restless Temper . . . ," *American Historical Review*, LXIX (July 1964), 969–989.

43. Turner Papers.

44. "Draft on League of Nations 1918—November," *ibid.*

45. "Turner, F. J. City, Frontier & Section; or, the Significance of the City in Amer. Hist.," *ibid.*

46. Billington, *American Frontier Heritage* (text edition) contains an annotated bibliography on the criticism of Turner's writings.

AFTERWORD

IN SEARCH OF THE FRONTIER

A Personal Adventure

Wilbur R. Jacobs

The Trailhead

I was born in Chicago, Illinois, on June 30, 1918. At the time Woodrow Wilson, a family hero, was waging a campaign for world peace and acceptance of the League of Nations. As I look back, it is not altogether surprising that Wilson's arguments for peace among nations would so deeply influence me in later years.

My forebears from Ireland and Germany seldom held back their antiwar sentiments. I remember how my mother's mother, born in County Cork, Ireland, smiled when she explained her unique blend of pacifism, vegetarianism, and Roman Catholicism. Her husband, an ironmaster who left militaristic Germany to establish a blacksmith shop in Chicago, agreed with her pacifism, but he was a husky atheist who liked meat, potatoes, and gravy. I adored my grandfather because he was a kind, generous man who gave me a silver dollar whenever I visited his shop. Here I saw brilliant sparks fly as he hammered and molded stubborn pieces of iron. I never knew my father's parents, but they, too, had no love for German militarism. In addition, they had another reason to leave Germany for America: my father's mother was Jewish and had suffered anti-Semitic slurs.

My father met my mother in Chicago. A company telephone secretary, she was a beautiful auburn-haired young woman who became a minor celebrity after her portrait appeared in a newspaper account of how she stood by her post during a dangerous fire. My father, managing a Jacobs teamster and real estate business, was financially responsible and handsome. He not only admired my mother but also shared her ideals about world peace. Although I'm certain that my parents were a romantic couple, I saw them as a stable, loving, church-going married couple with my sister and I. We lived quietly in a German neighborhood of north Chicago where I began my creep toward boyhood.

I have always believed that my father, Walter Ripley Jacobs, and my mother, Nona Isabelle Jacobs, both writers on nonviolence, were looking over my shoulder throughout my life. My mother, a Christian Scientist who often spoke about morality and the inner mind, was a lighted torch in the California antivivisection movement after we moved to the West Coast in the late 1920s. Tirelessly hammering away at her typewriter, she described, in pamphlets, reports, and other literature, cruel "scientific" experiments that brutalized cats, dogs, and other unfortunate animals. Her passionate correspondence with prominent individuals established her leadership in the burgeoning California animal-rights field. Her election as president of the Pasadena antivivisection society was no fluke or accident.

I identified with my mother and shared my close confidences with her. In college and later in the military, we conducted an extensive heartfelt correspondence. Her affectionately phrased missives expressed warmth, compassion, and style.

As a schoolboy I read dozens of public letters my father wrote to the press on domestic and international affairs and on animal rights. What most impressed me was his persuasive, elegant expression. Our family was elated when his photograph was published with an article proclaiming his election to the presidency of the Pasadena Humane Society. At home I heard dinner-table talk on humane society issues from a number of distinguished guests, some of them leaders in professions and politics. This experience inspired me in later years to accept the arduous commission as a national board member of the Humane Society of the United States.

My parents provided a warm, loving home for me and my younger sister, Shirley. As small children, we enjoyed an abundance of toys, pets, and competitive educational games. Mom and Dad encouraged us to proclaim our opinions. Praise was heaped upon me when I stood on a chair before relatives to give orations based upon my elocution lessons in rhetoric. One of the exercises, entitled "When My Pa Was a Boy," brought repeated applause. My rise in self-esteem was further stimulated by winning a neighborhood prize in marble shooting.

What I wanted most in life at this time was to be some kind of clone of Dad, a formidable, commanding masculine role model. His muscular figure easily absorbed a hard day's labor on desolate real-estate properties, back-breaking work that left me more than exhausted. In the evenings we had give-and-take talks about what Plato wrote or what Napoleon tried to do. For years Dad researched, although he never finished, a biographical study on Arthur Schopenhauer, the nineteenth-century German philosopher who wrote on animal abuse.

My family always lived in comfortable middle-class homes. I recall the beauty of our first brick bungalow on Central Park Avenue on the north side of Chicago. Above all, we looked forward to spending summer vacations at our delightful cottage by a lake at Wauconda, Illinois, some forty miles northwest of Chicago. While my father tinkered with repairs on our cottage, my sister and I turned outdoors, becoming expert swimmers in a short time. At a nearby dairy farm, we had lessons in the kindly care of horses, cows, hogs, and chickens. We always came home with fresh milk and eggs. I remember the hot summer day when I swam across our mile-wide lake, accompanied by my sister in a rowboat. Instead of garnering praise for this physical feat, I was scolded for my risky behavior.

One evening, my parents announced that we would be moving to California. My father had joined his younger brother in partial ownership of an orange grove in southern California. Everyone was overjoyed at the prospect of the westward journey, but we could not move in one trip. In fact, my family eventually made four successive automobile expeditions in the years 1927-1932.

In many respects, our journeys resembled each other. Once on the road, especially in the first several trips, my family banded together with other travelers into protective caravans of four or five cars. We helped each other when there were delays on bumpy dirt detours. Chuckholes menaced drivers and automobiles even on the main roads whether they were cement, asphalt, or gravel. Visibility ahead was oftentimes tricky because of dips, hills, or sharp flat curves. Everyone feared the wind and downpours of midwestern thunderstorms; tire chains were necessary to navigate muddy ruts. Yet, in all the trips, my family never suffered an

On the Road to California, ca. 1928

Automobile travelers caravaned for safety. Walter R. Jacobs is fixing a flat on the family Studebaker. Wilbur R. Jacobs is holding the dog's leash, his sister Shirley stands next to him.

(Photograph courtesy Wilbur R. Jacobs.)

The Jacobs Family Ford V8 Moving Slowly through a Sheep Herd on a Nebraska Dirt Highway, ca. 1932.

(Photograph courtesy Wilbur R. Jacobs.)

Wilbur, Walter, and Shirley Jacobs Standing with a Navajo Trader at His Post in Arizona, ca. 1928.
(Photograph courtesy Wilbur R. Jacobs.)

accident. Dad did all the driving and all that he could to avoid unsafe, as he supposed, "women drivers." After we completed our last trip, Mom learned to take the wheel—but not from her husband. She taught me to drive when I was thirteen.

Occasionally on lonely roads, Dad allowed me to steer our cars. At rest stops I helped him pencil out maps with alternative western routes. One northerly route, later called the Lincoln Highway, coursed through parts of Kansas, Nebraska, and Utah, following parts of the old Oregon Trail. Our favorite, however, was a southern one over a complex of highways approximating Route 66. We looked forward to familiar stopovers in old-fashioned towns such as Van Horn, Texas, Globe, Arizona, and Barstow, California. I recall scores of roadway "cabins," ancestors of today's motels, in which we generally passed each night. My father always checked these rustic shelters for unwanted creeping or crawling wildlife. On the road we trekked along with tire-repair kits, car tools, and jacks, together with fruit and bountiful sandwich supplies. For security we roped multiple suitcases to our car's running board. My parents were exceedingly tolerant of our dog, a scrappy fox terrier who often growled at my pet white rat. I was proud of how my parents scolded owners of bizarre roadside zoos, where wildcats, coyotes, and bears were captives in dirty small cages, sometimes without food or water, for tourists and travelers to gawk at. I felt a special compassion for tarantulas, lizards, scorpions, and snakes lodged in small boxes.

In the twenties and thirties, highways wound through the middle of small towns. Once on Main Street, Shirley and I seized the opportunity to plea for ice cream sodas sold at local drugstore fountains. On the highway she and I held contests in counting grazing animals or in replying to humorous messages on Burma Shave roadside signs. As the countryside rolled by, I liked to estimate the stages of ripening in cornfields and sometimes drew pictures of dissimilar red barns, some of them attached to houses. My mother always boasted that my farm pictures, especially of horses and black-and-white holstein cattle, showed genuine talent.

Emerging from the farm country of the Midwest, we were sometimes unaware of the gradual, almost imperceptible disappearance of trees and pasture as we drove onto the prairie ranchlands. But there was no way to overlook the stark beauty of the southwestern desert landscapes. At night, our automobile convoys ran along with beaming headlights through the Mojave Desert to avoid the daytime furnace heat. I fondly recall our mounting Cajon Pass and then the thrilling descent into the fresh, succulent greenery of southern California. The cool, moist air, the fragrance of orange groves, and the sheer magnificence of the mountainsides combined to give us a sense of triumph and delight. We made it to the golden land! Once in southern California we took family trips on the big red electric trains. What fun it was to clamber up the steps of those huge trolleys and be carried over the scenic rail lines of southern California. These were old-fashioned thrill rides for Shirley and me.

After the school year passed, we returned to the Chicago area in the early summer and then drove back again to the West in the fall. With each return trip to southern California, my family graduated to finer residences—from rental cottages in Alhambra to a comfortable Altadena stucco on north Allen Avenue near excellent schools and the magisterial Huntington Library, where I was to study in later years. For overland trips we invested in better cars each time, beginning with Jewett and Studebaker sedans and then moving on to a stately Graham Paige and later a zippy Model A Ford. Individually these cars had their own personalities. The Graham Paige, for instance, boasted power and beauty but coughed out and died with vapor lock in desert heat. The Model A

The Jacobs Family Posing by Its Graham Paige Sedan
This picture was taken in front of the Alhambra Cottage on Easter Sunday, 1930. Left to right, *Wilbur, Shirley, Nona, and Walter.*
(Photograph courtesy Wilbur R. Jacobs.)

was a superb road car but sported a small cab and offered little storage space. All rudimentary problems were solved by purchase of a sleek, black Ford V8 sedan in the 1930s. In case of repairs we could easily find a local Ford agency.

My father's proficiency as a breadwinner eased the Jacobs family through the bitter depression years. He showed me the social and financial benefits of recycling and refinancing residential real estate as well as the urgency of pursuing codes of business ethics. I soon came to appreciate his conviction that, beyond making money, there was a "higher calling." By his mid-forties, Dad gave virtually all of his time to public causes. He had attended Shurtleff, a Baptist liberal-arts college in Alton, Illinois, where he discovered his talents in music. Afterward, however, he concentrated on the law and the ministry as future careers. For health reasons, he said, he rejected those professions for an outdoor career in real-estate brokering, which gave him time to gorge himself on history and philosophy. A voracious reader with an encyclopedic memory, Dad lined the walls of our home with hundreds of volumes on history and philosophy, along with books for his children.

I became a fan of Edgar Rice Burroughs, Mark Twain, and Jack London. As a young adult, I was seduced by writings of a favorite of my father's, Bertrand Russell. Among the Russell volumes on our shelves were his *An Outline of Philosophy* and his subsequent *A History of Western Philosophy.* I was privileged to hear Russell's lectures (then considered controversial) at the University of California at Los Angeles (UCLA) in the late 1930s and will never forget the picture he cut—his white hair flowing and his pipe trailing smoke—as he emerged from Royce Hall classes. After brief conversations with him (I can't recall what was said) his sharp, ebullient gaze locked itself into my memory. I was more than glad to have been one of his defenders against campus bluenoses who nagged him for cohabiting with a younger woman. Despite Russell's early elitism, I applauded his positions on sex and marriage and his leadership in the "ban the bomb" movement.

School helped me to keep up with the book stacks at home. In both Chicago and Pasadena my education benefited from able teachers. I graduated from Pasadena City College (PCC) in 1938, after winning a varsity letter in a distance race called "the 660," but my ambitions were nonathletic when I entered UCLA. I had an inordinate hunger for reading books and journals and remembering what I read. Not surprisingly, bookworm that I was, I earned honors in both B.A. and M.A. degrees before Pearl Harbor.

My ambition was to become a history professor, a career objective that I outlined in a freshman college essay. Proudly independent, I became self-supporting. For instance, as a college student, I assisted Dad with repairing "fixer" houses. At UCLA I labored in Westwood department stores on Friday nights and Saturdays. I sold shoes, plastics, and childrens' clothing. At other times I worked in real estate and managed to teach briefly at UCLA and at Montebello High School, where I was an instructor in art and in mechanical drawing (keeping one day ahead of my pupils in night school).[1] None of these jobs, however, prepared me for military service.

The outbreak of World War II placed my conscience in a bind. On the one hand, I wanted to observe my pacifism; on the other, I equally wanted to serve my country against the forces of fascism and militant nationalism. At first I compromised by joining the army as a noncombatant soldier after the Japanese attack on Pearl Harbor. Although I promptly qualified as a trained medical ambulance driver, the military had other uses for my services.

When the time came to develop a mind-bending reeducation program for Nazi prisoners of war captured in North Africa, the army ordered me to assist in the task. As if it were yesterday, I vividly recall in late 1943 entering the heavily guarded gates of a prisoner of war stockade housing the cream of Hitler's Africa Korps. The stockade near Fort Sam Houston, a major air base not far from San Antonio, Texas, rose from dusty, hot flatlands like a small fortified city. As part of an educational Americanization program for Nazi troops, I lectured to audiences, seemingly attentive, on such topics as checks and balances in the United States Constitution, Jeffersonian republicanism, and the Turnerian theory of frontier democracy. I recall that my German students were fascinated with the germ theory of the Teutonic origins of United States governmental institutions. Aided by an interpreter (I commanded only fractured German), I gave talks to groups of prisoners in a small auditorium and shared their daily life. There was, I thought, a kind of bonding with some of the prisoners although their officers, in a barrage of crazy talk, declared that Hitler had bombed New York and that soon the Germans would rule the world.

In due time, however, our intelligence staff detected brutal murders by the Nazi *SS* officers. In nighttime secrecy, these cold-blooded Nazi maniacs, who had sworn their loyalty to their Führer, murdered some of their own soldiers whom they despised as "non-socialists" (i.e. non-Nazis). Informants told us that angry SS officers might even poison our food. I recall staring fearfully at mustard and catsup bottles on the mess-hall table.[2] Armed U.S. military police immediately separated Nazi officers from their men. Ordinary soldiers were parceled out to small staging areas; some were sent to agricultural labor camps. In the aftermath I received a letter of commendation for my part in undertaking Nazi reeducation.

The *SS* officers' murders of German soldiers stunned me, graphically demonstrating how little value the Nazis put on any human life. Consequently, I altered my status to that of full combat soldier. I would fight these Nazi madmen to the end of time to stop them, and I was under orders to join troops invading Europe, but the end of World War II came first. I had interminable thoughts about my encounter with the Nazis: one was that I resolved to devote my historical writing to opposing human injustice. As I was to find, however, to desire justice was to wish for something that can never be fully requited.

Renewed Academic Trails

After nearly four years of military service, there were rich opportunities for reentering academic life. Flattered to receive a Johns Hopkins University scholarship, I packed myself off to Baltimore to begin advanced studies in history. The Hopkins faculty, however, showed little enthusiasm for American frontier history. As a consequence, I departed for the West to accept a teaching invitation at my alma mater, PCC, and to enter a doctoral program at UCLA. Here, western history and Turnerian frontier scholarship were specialties of Professors Lewis Knott Koontz and John Walton Caughey, both reputable scholar-teachers and editors of the *Pacific Historical Review*. As an assistant in review offices, I had a chance to hone my writing skills.

At UCLA I easily located the person who would lead me on the path of American frontier history. Bespectacled Professor Koontz, a family friend, had taught UCLA extension classes, several of which my father took. When I entered UCLA, Koontz became my adviser. Sitting in his classes, I was agog at his lectures on the "unknown" George Washington in the frontier wilderness, his encounters with Indian people, and his treatment of African Americans. Koontz, as my doctoral mentor, guided my early research into Indian gift giving and wilderness diplomacy on the colonial frontier. Moreover, I won the post as his student assistant and quickly learned that I had a dual assignment as co-advisor for the UCLA Negro Club. Of poor-white southern background, Koontz was a dauntless leader-adviser for black students. During the 1940s I recall no African American faculty at UCLA to assist him. From Koontz I learned to appreciate the value of the little-known field of African American history. I was glad when his black students became my friends. Among them was Jackie Robinson, a high-spirited, scholar-athlete of immense talent and member of the UCLA Negro Club.

What awed me was that every field of study in the university boasted foremost experts, academic champions in their own way. When Koontz was a dinner guest at our home, my family discovered that he knew the eminent Frederick Jackson Turner and had arranged for Turner to talk before the Pacific Coast Branch of the American Historical Association when it first formed. What is more noteworthy is that my father had known Turner as a fellow member of the Pasadena Neighborhood Church, a Unitarian congregation that also included the famed Robert Millikan, founding father of the California Institute of Technology. As a young man, I was thoroughly Turnerized, but my captivation with the master's theories had pluses and minuses. Ten years would pass before I emerged from the Turnerian umbrella to see western history in a more balanced perspective, especially in Indian and environmental studies.[3] Yet the remarkable saga of Turner's ascent and partial decline did not prevent me from appreciating the value of his frontier and sectional theories, which are still hotly debated today by modern critics.[4] Turner, limned as he was by criticism, nevertheless became a focus of my future studies.

Early in my teaching career at PCC and at Stanford, I published a book and articles on Turnerian themes and Indian-white relations. Bubbling over with fervor, I excitedly told my students storylike research adventures about historical figures. My work so captivated me that in the classroom I practiced Sherlock Holmes techniques in historical detection. Teaching was fun, and students responded with enthusiasm.

More and more I came to respect the personal value of individual pupils. Perhaps I overdid my office hours, talking at length with students about their career plans. Yet there were immense rewards. Through such dialogue I learned how to document the kind of student recommendations that brought results. As a temporary instructor, I earned recognition for excellence in teaching at Stanford, where I conducted "Western Civ." Although a stimulating freshman bread-and-butter course, it had one drawback: it was based upon the thinking of dead white men.

By 1950 I was an experienced college instructor with high approval ratings, but it was uncertain that I could win any key position over other well-qualified applicants, particularly those trained at prestigious eastern universities. However, my efforts were soon rewarded. The prize was a tenure-track post at the new branch of the University of California at Santa Barbara (UCSB). After months of waiting I finally received confirmation that I had actually been selected for this new "publish or perish" position. I remained at UCSB for the rest of my teaching career. At various times I taught everything from political science and government to the history of California, early America, the West, and American Indians.

In addition I took on lecture assignments in general American history and the history of western civilization.

While the schedule was demanding, I tried to relate teaching to my fields of scholarship. Notwithstanding my pride in being prepared for meeting my classes, I gave more than equal time to research. In fact, I seldom began lecture preparations before nine P.M. in the evening, devoting the major part of the day to research and writing. My intent was to activate a dynamic research program. I found great pleasure in unravelling stories of the early frontier. Moreover, I believe that I was on the trail of truth. I remember how proud I was when legendary historian Bernard DeVoto verbally complimented my efforts to trace Parkman's Oregon Trail. In the end I became an enthusiastic workaholic, steadily turning out book and journal publications. To my astonishment, one of my essays, footnotes and all, was completely plagiarized by a novice scholar from a small college.

My studies in American Indian history led me to comparative analysis. I gave exhaustive effort to combing libraries for source materials wherever I was teaching, whether at UCLA, the University of California at Berkeley, the University of Colorado, and other institutions. During visits to homelands of native peoples, I gathered oral history accounts, and in libraries around the world I examined evidence that indigenous peoples were swindled and brutalized by colonial profiteers, missionaries, and their offspring.

During my research I learned that, while swindling was going on, something else had happened: hundreds of thousands of native women and European and Anglo American men had intermarried. These interracial unions (reinforced by occasional intermarriage of European women with native men) created large Metis populations in the United States and Canada and a rapidly multiplying southwestern *mestizo* populations in the Southwest (a blending of races often called the *mestizaje* process). Through my editorial research for the Smithsonian Institution volumes called the *Handbook of the North American Indians,* I was moved by the history of the Seminoles, a mixture of Native American, African American, and Spanish peoples, who fought heavily armed American troops in a series of punishing jungle wars in Florida during the first half of the nineteenth century.

Given my parents' work in behalf of world peace, I am proud that my investigations helped spotlight the unique phenomenon of peaceful accommodation on certain frontiers of western America. From the very beginning of frontier history, native American women and men, often unsung promoters of peace, were part of an ignored story of nonaggressive Indian-white relations.[5] Yet pioneers, often vehemently anti-Indian in their memoirs, passed on their hatred to historians.

By contrast, I was delighted to read in Quaker and Moravian records many reminders of Anglo-Indian peace.[6] When William Penn welcomed Delaware people with "I love you," I am convinced that he meant every word and that the Indians believed him. In the entire history of colonial Pennsylvania, no Indian wars erupted until the Quakers lost control of the provincial Assembly in the late 1750s. The long peace in Pennsylvania is convincing evidence of the generous gift-giving between the Quakers and their Indian neighbors. The model Quaker farms among the Iroquois demonstrated sober habits of work and productivity—some of the best white behavior found on the American frontier.

And Moravian missionaries, sometimes called German Quakers, were similar in outlook, especially John Heckewelder, who left a poignant account of his life among the woodland tribes. Moravians provided sanctuary for tribal people at a Conestoga, Pennsylvania mission. After losing their families to Indian attacks, the nearby Paxton Boys, encouraged by their Presbyterian ministers, turned their rage on and murdered these peaceful Indians. This unpleasant story, bursting from warfare in Pontiac's Uprising, illustrates the paradox of writing about triumphs and disasters. For all the spilled blood in this episode, there was a certain muddleheadedness in our understanding of Pennsylvania in the early 1760s. To clarify the issues I dove into the archival records and wrote a booklet, *The Paxton Riots and the Frontier Theory.*[7] My conclusion had a familiar ring: violence begets yet more intensive violence, and nobody wins.

The harsh, underside of native-white relations is clearly evident in Australia. During the 1960s in Queensland, on Palm Island off the coast, I interviewed Aborigines actually locked up for decades in a huge insular government prison for no reason other than they were native people. When I made a public protest about penitentiary conditions, my picture appeared on the front page of the *Canberra Times,* accompanying an article, "American Professor Exposes Prison in Paradise."

Later, at Australia's New England University I met a graduate student who documented systematic poisoning of Aborigines at sheep "stations" (i.e., ranches) on Queensland's frontiers. The student's adviser, an internationally known historian, confirmed the accuracy of the poisonings. This kind of subject matter, I found, was unpopular in mainstream and official Australian history.

At this time I wanted to adapt the techniques of comparative ethnologists.[8] The first step came when I had a chance to leave the library and embark on an expedition into interior Sepic River villages of Aborigines in Papua, New Guinea. With the aid of a native guide, a Seventh Day Adventist medical missionary, I lived for brief periods at jungle villages, where I met men and women who, although having recently given up headhunting, lived much as their ancestors did thousands of years ago. Warriors still held sway in this society. I was depressed to observe the near slavelike status of native women (called Marys) who seemingly worked night and day to care for broods of children and to salvage a kind of white dough from trunks of the Sago palm. My guide astonished me with the assertion that should any Mary wander into the men's sacred "House Tamberain," where men spent long hours talking over sacred objects before a ritual fire, she would "expire." When I asked for clarification, my guide shouted, "Yes, she die. I see her die."

I had respect for the village warriors, who, with powers of enchantment and force, protected their homelands from conquest by other villages or by Europeans. On the morning after arriving in "Swago" village, which was perched on a small rise straddling a tributary of the Sepic River, I awkwardly climbed down a wobbling log stepladder from a palm-covered cabin sitting on pilings about ten feet off the ground to isolate us from hungry insects. A muscular headman-warrior was slashing grass beneath our hut with a swordlike knife. Seeking to begin a dialogue with him, I asked in pidgin English, "Headman Moonkey" (moonkey is a racist word for a native man in South Pacific Island pidgin), "what do you like to eat?" He gazed at me a moment and replied in correct Oxford English, "I eat anything that moves!" One can be sure I froze after this reply; the headman, it turned out, had resided at a Lutheran mission school. He proved to be extremely cordial and introduced me to families who crafted weapons and decorative masks. He also taught me about village social codes and clued me to layers of expressive meaning found in pidgin English.

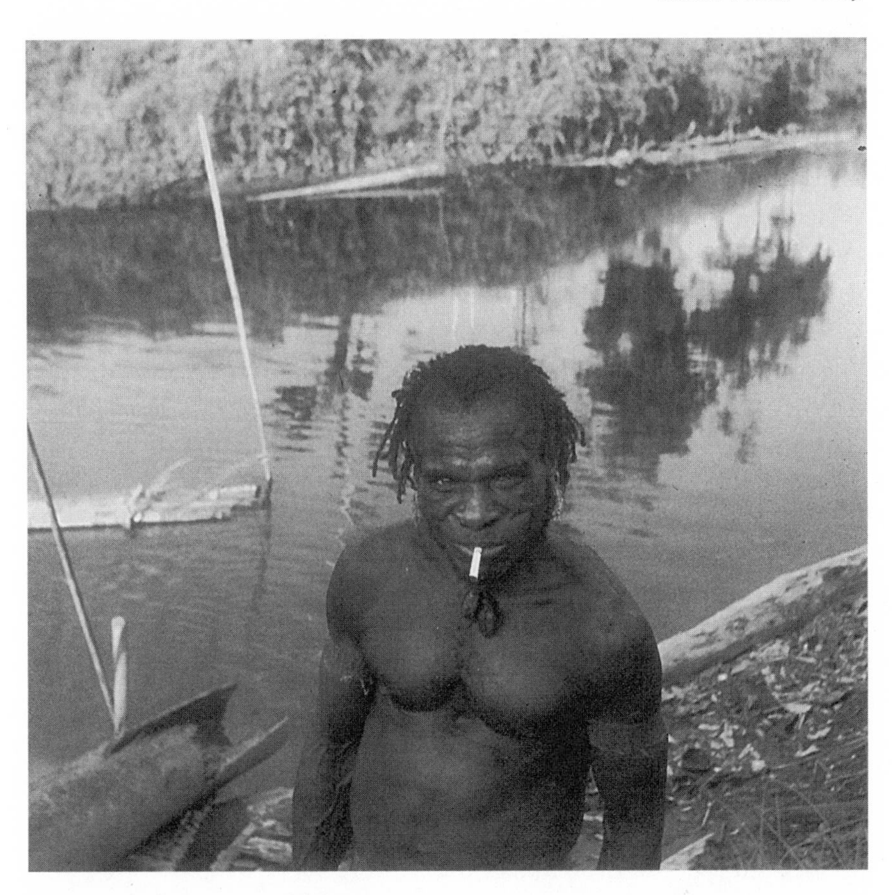

Sepic River Headman at Sago Village

This strong individual, whose immediate ancestors were headhunters, became Jacobs' friend, guiding him through the village and explaining subsistence patterns and the roles of men and women.

(Photograph courtesy Wilbur R. Jacobs.)

There was a lesson in this kind of hidden meaning when I listened to a "singsong" on the eve of my departure from a village's mud landing bank. I was saluted with a vibrant melody sung with the words, "He Old Fella, Good Fella." I was the "Old Fella" since their life span was about forty years,[9] and I was a "Good Fella" because I had paid Australian dollars for several decorative masks. It was clear that I violated a standard trading-goods schedule that paid much less for masks and weapons and cheated the craftsmen of this Sepic village. My guide had frantically called to me, "No pay dolla, pay quarta!" The experience reminded me of the dramatic play, "Ponteach," by an American colonial frontiersman. At one point in the drama, a trader scolds another for his honesty: "It's no crime to cheat and gull an Indian." My scruples told me that these impressive New Guinea masks were worth more than any Australian dollar, and any sin I had committed was underpayment. But these were the ways of wilderness trade and part of the give and take between indigenous peoples and Europeans. Here was a key to comprehending much of the double-standards of justice in American fur-trade history.

After returning from this somewhat hazardous expedition, a physician told me I was a lucky man. My medication at the time saved me from malaria-carrying mosquitoes that can cover the unprotected skin with a buzzing, brown, infectious blanket. As much as I admired the Sepic men and women who fought off European conquerors, I decided that malaria was probably their best defense against white invaders.

Trails of Ethno-Environmental History

Subsequently, after returning to America, alone or with advanced students, I experimented with other kinds of investigative research beyond the library. We camped out at Indian reservations and villages in Canada, Mexico, and the United States. We interviewed traditional tribal peoples and poured over Indian oral history tapes in the Doris Duke Collection at the University of Utah Western History Center. I conferred with tribal craftsmen and bought samples of their unique pottery at the old Catawba Reservation in Virginia. After participating in rituals at the Six Nations Long House at Onondaga, New York, I talked about the injustices in the Fort Stanwix Treaty of 1768. At another time I pranced in Iroquois war dances at their reservation in Brantford, Ontario. With a former student, I joyfully observed rain dances and thunder showers that followed at Hopi Second Mesa.

Our visits were not always welcomed. Certain Navajo trustees at the Rough Rock School were quietly resentful of our presence, and one of them was reportedly a witch. I remember his refusal to speak to me in English when I tried to explain my stake in Indian education at UCSB. At the Wounded Knee trials, however, Indian leaders, including Russell Means, Dennis Banks, and Vine Deloria, welcomed me when I testified on the terms of the Sioux Treaty of 1868. During the trials I came to

Old Posey, One of the Last Ute War Chiefs
Leading his people against the Utah militia, Old Posey was murdered by arsenic poisoning, according to Indian oral history tapes in the Doris Duke Collection. (Photograph courtesy Wilbur R. Jacobs.)

know excellent lawyers, who taught me more than a little about the history of Indian law.

Despite differences in detail, my accumulated evidence consistently revealed a history of depraved Anglo behavior toward Indians. For instance, older Indians repeatedly told in oral history accounts about the murder of Ute chief Old Posey with arsenic-laced bread given to him by Mormon agents. San Juan County western historians vehemently rejected this version of Old Posey's death, claiming that he had been killed in battle.

No Anglo American is proud of the germ warfare that killed native peoples. My investigations helped confirm the record of such malevolent practices. We can thank the middle-aged historian Francis Parkman for first bringing the premeditated spread of smallpox to our attention. After painstakingly verifying text and notes for a second edition of *The Conspiracy of Pontiac,* he came upon incriminating letters of a mean-spirited general, who ordered that "gifts" of tainted smallpox blankets be given to "savages" surrounding Fort Pitt during Pontiac's Uprising. That Parkman was disgusted with what he uncovered is an understatement.[10]

In my own experience this kind of incident illustrates a persistent theme in western American history: the portrayal of Indians as brutish survivors of a savage past that should be erased from the memory. From this perspective America can be seen as nation with an upside-down frontier history that pictures dispossessed, starving Indians as vicious predators who literally needed to be pushed out of the way, as a noted geographer has argued. Old-style writers of the Walter Prescott Webb school and others who followed Turner told tales of "savages" and their "depredations."

After hundreds of research hours, I collected enough data to show that able Indian pacesetters made policy and created solutions as they negotiated treaties. The chiefs, aided by woman advisors, painstakingly considered tribal interest and avoidance of future conflicts in statesmanlike speeches, in proposals, and in counterproposals. I was more than disappointed to discover that the United States government ignored such facts and, in a booklet, depicted Indian leaders as simplistic, happy tribesmen who willingly gave up their landed heritage.[11]

The more I considered Indian history, the more I saw its challenges. Although I studied the subject for decades, almost every week I discovered new facts, partly as a result of my friendships with individual Indians. Grandfather, an amiable, wise medicine man at Redwind, an Indian commune near Santa Margarita, California, became a friend and counselor. I remember that, during a visit with students to his encampment, he gave me a handmade abalone necklace that he had blessed in the Redwind kiva. "It will keep the bad guys away," he told me. When I smiled, he said, "Try it out." I can tell you that I still have a feeling of security when I wear the necklace, and I swear that it helps fend off hostile reviewers of my books.

Besides knowing Grandfather and other medicine men, I was blessed with Indian students, many of whom were also friends. Some later became dentists, lawyers, or professors. I have always been extremely proud of their scholastic and professional accomplishments. Because I was the UCSB advisor for Indian students, I knew their times of joy and sadness. I went to powwows, weddings, and funerals. Serving as a human resource to Native American students, I tried to share with them my experience in American universities and the Anglo world and to help guide them past the pitfalls to the benefits and rewards. There were, then, clusters of contacts with Indian friends that gave me confidence in my own perceptions and cultivated my appreciation of cultural diversity.

As my learning curve ascended, I managed to control my tendency toward polite brashness. My journals, re-enforcing memories of long forgotten details, show how easily I was roused to quiet indignation. Over the years I found that the issue of Indian removal was continually clothed in factual bias and controversy. In my classrooms, students heatedly debated the question. Afterward they carried arguments into the hallways and then into the university courtyards. Needless to say, this kind of student debate influenced my own investigations and writings.

My zigzag path in writing for publication began when, as a UCLA senior, I wrote about gift-giving power struggles between Indians and whites on the eighteenth-century frontier. I was elated when my book, *Diplomacy and Indian Gifts,* was awarded a prize by the Pacific Coast Branch of the American Historical Association.[12] Not unexpectedly, I eagerly embarked on further explorations of what blossomed into a new field, ethnohistory. In 1960 Samuel Eliot Morison wrote an elegant front-page essay in the *New York Times Book Review* commending my two-volume edition of the *Letters of Francis Parkman.* The book, I was told, was a finalist for the Pulitzer Prize and was subsequently chosen for the John F. Kennedy White House Library collection.

Although the 1960s brought student riots, bombs, arson, and even murder to UCSB, there were surprisingly few reverberations in the history department, which had a well-deserved reputation for conservatism. Against entrenched opposition I fought to sustain the study of Indian history and supported Roderick Nash's struggle to establish environmental studies. Additionally I helped create a program in religious studies and backed a new course in women's history. At the same time I carried on a guerrilla campaign to fortify a course in African American history and a program to win scholarships for black students.[13]

But I devoted most of my energies to my fledgling Indian history course, History 179Q. Unexpectedly my chair told me to add the letter *Q* because Indian history could not be counted as an alternative history class for an "American institutions" graduation requirement. By the 1970s, after changes in the power structure of the history department, a progressive wind swept through to clear the stale, musty air.

When my Indian history course finally appeared in the university class schedules, it drew students from all majors and set a precedent for other programs in minority history. Equally significant, History 179Q proved to be the first American Indian history course offered on any campus in the University of California system.

At the time, in the 1970-1980 era, History 179Q blazed a trail to senior and graduate seminars. A series of doctoral candidates wrote prize-winning books on Indian history.[14] These students brought an agglomerate of controversial findings to the seminar table. When I set forth my own views in public lectures and essays, I was at first confounded to find that I occasionally encountered quiet reluctance to accept my interpretations.

I believed that I had the facts and that I took a neutral stance. Yet some of my advanced students generated counterarguments. As one of my able pupils (who co-authored a controversial paper on Father Serra's Indian policies, which we prepared for a conference of the American Society of Ethnohistory) said, "Will, you look at these documents one way, and when I study them I come to another conclusion." We debated Father Serra's endorsement of the "Just War" theory, which had been used to explain away the burning of Indian villages and the murder of survivors. We read and analyzed the data and decided that we would let our readers make their own judgments. After several years we are still writing and rewriting![15] Part of our work on Father Serra was based upon the writings of my friend, Sherburne F. Cook, a role model and talented scholar who contradicted rosy misconceptions of California mission-Indian history. Soon after his death, when he was under bitter attack by Franciscan historians who resented his treatises sharply critical of the treatment of native people in California missions, I wrote an essay to defend his noteworthy scholarship.[16]

Following my retirement in 1988, Indian history at UCSB floundered. Despite the success and popularity of the field, the department failed to hire a replacement.[17] In spite of this disappointment, I was gratified to see that my work received increasing acceptance among fellow historians. And I instantly felt sympathy for the innovative studies of New Western Historians, having trudged through the same mud over some of the same trails.

Legendary Mentors

In retrospect, my writing improved when I tuned it to the higher prose tones of legendary historians Francis Parkman and Frederick Jackson Turner. To upgrade my writing craft I devoured everything that came from the pens of these notable writer-scholars. This done, I moved on to ferret out their methodologies as they could be gleaned from their letters, diaries, and research materials. There were lessons to be learned in Turner's scheme of massing materials in a carefully organized database with thousands of three-by-five cards. Like Turner, who was overwhelmed by his data, I made out thousands of note cards and was equally overwhelmed. Like Turner, I tried to skim the best from an ocean of facts, but it was hard to compete with the old master.

Parkman was different. His references were his journals and lengthy copies of documents that he had read aloud to him. He was partly blind and took few notes. Time and again he made journal comments on the need for "observation." In taking field notes he followed a policy of intense watchfulness.

On my part, I soon disciplined myself to keep research journals for preserving and scrutinizing minute but meaningful details when and wherever they appeared. But there were problems. What was important? My journals and boxes of research cards bulged with unusable materials. Again, I tried to follow Turner's example—to feel the way forward a little at a time. The ideal was to put my research findings into articles and then expand them into books.

My spearhead for research converged on two specialties merged as one: ethno-environmental history of the American frontier. When I studied the history of Indian homelands (culture area studies in anthropology), I found myself zeroing in on biospheres, biodiversity issues, ecological flux, all converging on American frontier expansion (or conquest). Major fruits of my research appeared in a recent book, *Francis Parkman, Historian as Hero: The Formative Years.* In that volume I analyzed events in Indian history that clarified Chief Pontiac's leadership. At the same time I assessed Parkman's remarkable skill at creating ethno-environmental perspectives in his narratives.

This kind of approach engendered for me stunning perspectives on western history. Let me cite another example. In one generation, after displacing Indians, Anglo-American pioneers occupied the area of the Louisiana Purchase and brought in a huge population of domestic farm animals. In the process they established a mighty agricultural basin, wiped out vast populations of wild hoofed animals, altered or decimated much of the original plant life, killed off thousands of tribal peoples, and confined their survivors to reservations. There was crying need, I concluded, to be aware of such ecological turning points in our history.

To understand better what had happened, I traveled by car and by foot from the Atlantic tidewater to the Appalachians and westward. With notebooks and tape recorders in hand I went to Indian reservations in eastern and western America, in Canada, and in Mexico.

Then there was a long paper trail of governmental reports and documents. These self-serving official documents could not always be trusted. A good antidote, I found, was comparing government documents with private reports, diaries, and letters. My data was scattered. When I pestered reference librarians at universities, they gave me a generous outpouring of help, but I'm sorry to report that, at certain private institutions, curators calmly declined to let me examine particular caches of source materials that were stashed away and reserved for favorite scholars.

My central path was along the meandering line of the Indian frontier, which retreated westward from the Appalachians to the Great Plains and eastward from the Pacific Coast over the Sierras into the Great Basin. The evidence revealed an uneven yet steady westward and eastward advance of Anglo-American peoples, a subject that is now under intense critical scrutiny by modern writers.[18] In pursuit of Turner's trail and merging it with my own findings, I encountered no barriers or distinct boundaries at the Missouri River as some historians have argued.[19] The evidence showed European conquests of Native Americans left turbulent displacements everywhere, a topic on spacial history involved in my investigations on America's environmental frontier.[20]

Turner, I have concluded, was intrinsically accurate when he described frontier processions of settlers treading through the Cumberland Gap and, a century later, through South Pass. But along with my Indian frontier studies, I soon took note of less obvious but parallel frontiers of encroachment on wildlife, plants, and natural resources.

To comprehend the mysteries surrounding the extraordinary disappearance of wildlife, I turned to the powerful words of the German Nobel Prize winner, Konrad Lorenz. It was as if I had entered a completely new world of learning, the science of ethology. In two books written for the layman, *King Solomon's Ring* and *Man Meets Dog,* Lorenz gave us a look at the fruits of his investigations on animals. As I moved on to study his scientific papers, I was convinced that his kind of observation of animals in natural settings revealed behavioral patterns that could not be learned in man-made settings of zoos or laboratories. As the astute Sally Carrighar revealed, Lorenz provided data for the understanding and appreciation of the our wildlife heritage.[21]

For the first time I understood how much a historian of the frontier could profit from knowledge of an individual animal's normal behavior in its natural habitat. What is significant in our history is that the habitats were systematically destroyed. I found that, without realizing the compass of their actions, ordinary settlers, marching westward, brought chaos to wildlife by merely fencing their fields and pastures. More deadly were their organized campaigns to kill off the animals. Huntington Library photographs, Mormon records (of John D. Lee's "War against the animals"), and John Muir's papers and writings told me parts of that sad story. After the passing of the frontier in the twentieth century, there is another dismal account that has yet to be fully told: killing off predators by use of the deadly poison, 1080.[22]

Besides John Muir and Konrad Lorenz I was stimulated by other insightful writers. For instance, Aldo Leopold, at one time a professional hunter, traced environmental despoliation to the careless desecration of wildlife in Wisconsin as well as in Arizona and New Mexico. I will never forget his admonition to "think like a mountain" when we gazed upon deer overbrowsing the plantlife on a mountain slope. Leopold then raised the question of why he was one of those who killed off wolves, one of nature's predators. The prescient Rachel Carson impressed me when she demonstrated that pesticides endangered whole species of birds throughout twentieth-century America at a time when the old western frontiers had disappeared. Now I faced questions of the interface between biodiversity, habitat preservation, and the story of progress in the westward trek. Should there be a historian's perspective of a "land ethic"? "Yes," I emphatically answered, but there were no easy answers to these complex issues at the intersection of environment and history.

During the 1980s, my last decade of teaching at UCSB, I concluded that I should give more attention to population issues and their relation to the occupation of western America. I had already plunged into Indian demography, publishing a 1974 *William and Mary Quarterly* article, "The Tip of the Iceberg," but I was disappointed to find that modern United States population history as a field of study had virtually disappeared.[23]

That practically any environmental issue had roots in population booms was troubling. After lengthy discussions and a television debate on population increases, pollution, and politics, I was confounded by the increasing coarseness of the dialog. Such controversial issues, I was reminded, reverberated out of the legacy of socialist-novelist Upton Sinclair. We now know that Sinclair was effectively accurate when he warned about how polluted runoff from Chicago's "Union Stockyards" threatened both people and the Chicago River. The city of Chicago was not the first urban center to face such pollution. In London of the 1850s, gushes of city sewage killed the fish in the Thames River. The cause was deoxygenation in a formerly healthy salmon river.[24]

As an observer of Chicago's meat industry, I had a particularly unpleasant experience. While monitoring Chicago slaughterhouses' use of humane food-animal slaughter guidelines for the Humane Society of the United States, I was struck by a consequential historical fact of life: contrary to popular belief, the cattle industry, east and west, was dominated by the corporate meatpackers, which owned and produced a lot more than ranches, drives, and cowboys. Feeding steers generated millions of gallons of waste. Silent streams of urine and liquid cow manure oozed everywhere around the slaughterhouses and, from there, washed into rivers, streams, and pools. From Cincinnati to Chicago, Omaha, and other points west, a poisonous trail was left by feed lots and meatpackers. The result: pollution now threatens the Ogallala Aquifer, the greatest source of pure water on the High Plains. Along with this ecological disaster, the water users are draining the aquifer to the extent that it may not be able to renew itself.[25] Such problems are augmented by midwestern factory hog farms that generate tons of manure-urine liquid, a witch's brew that is poured into neighborhood "lagoons." Again, how much of this negative stuff should be told in the story of the American frontier-West? This is a vexing question for our best historians.[26]

After receiving an appointment at the Huntington Library as research scholar in 1992, I had time for fresh investigation. Moreover, I enjoyed a new title and was no longer "over the hill" as an emeritus professor. My changed status, along with support from a wonderful young family, gave me renewed energy. I not only plunged into activities of Boy and Girl Scouts and soccer and Little League, but still contrived to write books and articles, to speak at the UCLA Indian Center, and to lecture in a new environmental history course designed by Scott Dewey at the California State University, Los Angeles. Additionally, I had the chance to taste the reality of helping the homeless and coping with other socioeconomic and environmental problems as board member of a nonprofit corporation, The Economic Round Table of Southern California. I also met kindred spirits on the board of Pasadena's Throop Memorial Unitarian Church.

In rereading my essays in this book, some written decades ago, I see more clearly how my personal life interplayed with my professional career as a historian of the American frontier-West. As my curiosity matured, I tended to lose interest in seeking personal and professional recognition for my historical work. My objective has always been to make America a better place to live. Having had the opportunity to travel in many parts of the world as a visiting professor, I have observed enough of human life in other nations to confirm my belief in America's greatness. I am supported in this belief by my family, William, Emily, and my wife Priscilla, together with my two older daughters, Betsy and Catherine, who have encouraged me to continue writing on the grand history of the American West.

NOTES

1. Later, at UCLA, as an art minor, I was told by one of my teachers, "Wilbur, you have not much talent." Nevertheless, I did overcome my limitations enough to teach high school art and mechanical drawing for a brief period.

2. For a different perspective, see Ron Robin, *The Barbed Wired College: Reeducating German POWS in the United States during World War II* (Princeton, 1949).

3. Richard Hofstatder, in talks with me at the Huntington Library in the mid–1960s on consensus writings, recalled his similar struggles to emerge from Vernon Parrington's shadow.

4. See Wilbur R. Jacobs, *On Turner's Trail: 100 Years of Writing Western History* (Lawrence, 1994), 203–247.

5. Yale historian John Mack Faragher spoke on these themes in a recent lecture, "The Great West and Greater America," Ray Allen Billington Lecture, Huntington Library, 15 March 1995.

6. As the Moravians protected their mission Indians, so did their neighbors, the Delawares, find sanctuary for the "brethren" during the horrors of the French and Indian War. Donald F. Durnbaugh, ed., *The Brethren in Colonial America* (Elgin, Ill., 1967), 162–163.

7. Wilbur R. Jacobs, *The Paxton Riots and the Frontier Theory,* Berkeley Series in American History (New York, 1967).

8. Francis Parkman and, later, William N. Fenton had been students of the seventeenth-century Jesuit, J. F. Lafitau, who compared Iroquois to tribal people of biblical antiquity. See Fenton's pathbreaking article, "J. F. Lafitau (1681–1746), Precursor of Scientific Anthropology," *Southwestern Journal of Anthropology* 25 (Summer 1969): 173–187. Fenton has also edited and translated Lafitau's *Moeurs des Sauvages.*

9. Leprosy was one of the dreaded diseases of these jungle peoples. I remember an emaciated woman, with one wasted stub of an arm and a deformed remaining hand, grabbing my fingers in a greeting after my guide had given her medication. "Can she pass on the disease?" I asked my guide, a medical missionary. "No, if she takes her medicine, but who knows if she takes it," he replied. I washed my hand furiously in nearby river water.

10. See Wilbur R. Jacobs, *Francis Parkman, Historian as Hero: The Formative Years* (Austin, 1994), 84 ff, 199–200.

11. The U.S. State Department, through its offices in Canberra, Australia, distributes a booklet reporting on so-called blessings America has bestowed upon Indians, who joyfully give up their land and smile (in accompanying pictures) to prove their happiness.

12. Wilbur R. Jacobs, *Diplomacy and Indian Gifts* (Stanford, Calif., 1950). The book was later reprinted by the University of Nebraska Press in the Bison series as *Wilderness Politics and Indian Gifts.*

13. One of these gifted African American students at UCSB was Otey M. Scruggs, who won a scholarship to Harvard and afterward taught history at Syracuse University.

14. Among those who distinguished themselves with pathbreaking published dissertations were Calvin Martin, Albert Hurtado, Yasu Kawashima, Christopher Miller, Gregory Schaaf, Robert Trennert, George Frakes, and Georgiana Nammack. I am especially proud of the William Swagerty's record as an editor and writer at the Darcy McNickle Indian Center at the Newberry Library and of the revisionist portrait of Roger Williams created by Robert Brunkow. George Frakes and Curtis Solberg published important studies in environmental history. Robert Righter earned a national reputation as a historian of national parks and wind energy.

15. Wilbur R. Jacobs, "Father Serra's Franciscan Indian Policy," with Jason Suarez, UCSB doctoral candidate (unpublished).

16. For my appraisal of Sherburne F. Cook, see the "Sherburne F. Cook," *Pacific Historical Review* 54 (May 1985), 191–199.

17. As of January 1995, Indian history at UCSB has a supporter in Professor Ann Plane, who began a revival of field with a renaissance course, History 179, offered this time without the "Q."

18. Wilbur R. Jacobs, *Dispossessing the American Indian*, 2nd ed. (Norman, 1985), xvff, 126–150.

19. See Richard White's distinguished book, *"It's Your Misfortune and None of My Own": A New History of the American West* (Norman, 1991). In his introduction, White argues, "The boundaries of the American West are a series of doors pretending to be walls. On the north and south the West ends in arbitrary lines drawn on the map. On the east and west real physical entities—the Missouri River and the Ocean—do mark its limits." *Ibid.*, 3.

20. My current research project is a book-length study concentrating on eyewitness accounts of environmental change on America's frontiers.

21. See for example, Carrighar's exemplary volume, *Wild Heritage* (Boston, 1965), especially her eulogy to Konrad Lorenz, pp. 34–36. Another eye-opening nature study is her delightful *One Day on Beetle Rock* (New York, 1956).

22. In early 1970, by executive order, Richard Nixon halted use of the deadly poison, 1080. I wrote to him, commending his order, and was honored to receive a cordial reply over his signature.

23. See Russell R. Menard's probing essay, "Whatever Happened to Early American Population History," *William and Mary Quarterly* 50 (April 1993), 356–393.

24. See the volume by a noted British scientist, Kenneth Mallanby, *The New Naturalist, Pesticides and Pollution* (London, 1967), 50–51.

25. The Ogallala is also known as the High Plains Aquifer. For analysis of the contamination by nitrogen fertilizers and feedlots, see David E. Kromm and Stephen E. White, *Groundwater Exploitation in the High Plains* (Lawrence, 1992), 52 ff. On the depletion issues, see ibid., p. 70 ff. On the western myth of the inexhaustibility of the aquifer, see John Opie's excellent volume, *Ogallala, Water for Dry Land* (Lincoln, 1993), 163–164. John Robbins, a leading nutritionist of America, has a superb analysis of the pollution of the High Plains in his *Diet for a New America* (Walpole, N.H., 1987), 370–371.

26. For an astute critique of the negative view of western history, see essays in Gene Gressley, ed., *Old West/New West* (Worland, Wyo., 1994), 28–49, 122–147.

SELECTED BIBLIOGRAPHY OF WILBUR R. JACOBS

N.b.: Within categories, books and articles are listed chronologically by original date of publication.

BOOKS (AUTHOR)

Wilderness Politics and Indian Gifts. 1950. Reprint, Lincoln: University of Nebraska Press, Bison Books, 1968. Originally titled *Diplomacy and Indian Gifts*.

The Historical World of Frederick Jackson Turner. New Haven: Yale University Press, 1968.

Dispossessing the American Indian. New York: Scribners, 1972.

Dispossessing the American Indian. 2nd ed. Norman: University of Oklahoma Press, 1985.

Francis Parkman, Historian as Hero, the Formative Years. American Studies Series. Austin: University of Texas Press, 1992.

On Turner's Trail: 100 Years of Western History. Lawrence: University Press of Kansas, 1994.

The Fatal Confrontation: Historical Studies of American Indians, Environment, and Historians. Historians of the Frontier and American West, ed. Richard W. Etulain. Albuquerque: University of New Mexico Press, 1996.

BOOKS (COAUTHOR)

Survey of American History. With Louis K. Koontz and Kenneth P. Bailey. Ann Arbor, Mich.: Edwards Bros., 1949.

Turner, Bolton and Webb, Three Historians of the Frontier. With John W. Caughey and Joe B. Frantz. Seattle: University of Washington Press, 1965.

BOOKS (EDITOR)

The Appalachian Indian Frontier. 1954. Reprint, Lincoln: University of Nebraska Press, Bison Books, 1967.

Letters of Francis Parkman. 2 vols. Norman: University of Oklahoma Press, 1960.

America's Great Frontiers and Sections. 1965. Reprint, Lincoln: University of Nebraska Press, Bison Books, 1969, and Landmark Edition, 1977. Originally titled *Frederick Jackson Turner's Legacy.*

The Paxton Riots and the Frontier Theory. Berkeley Series in American History. Chicago: Rand McNally, 1967.

Benjamin Franklin, Philosopher-Statesman or Materialist. New York: Holt, Rinehart, and Winston, 1972.

ARTICLES IN BOOKS

"Henry Chatillon." *The Mountain Men and the Fur Trade of the Far West*, edited by LeRoy R. Hafen, 229–31. Glendale, CA: The Arthur H. Clarke Company, 1965.

"Romance of History in America." *Intellectual History of America: Contemporary Essays on Puritanism, the Enlightenment, and Romanticism,* edited by Cushing Strout, 15–22. New York: Harper and Row, 1968.

"Suggestions for a 'New Look' at Frontier History." *The Training of Western Historians,* 275–80. Tucson: University of Arizona Press, 1968.

"British Colonial Attitudes and Policies Toward the Indian in American Colonies." *Attitudes of the Colonial Powers Toward the American Indian,* edited by Howard H. Peckham, 81–106. Salt Lake City: University of Utah Press, 1969.

"Cadwallader Colden's Noble Iroquois Savages." *The Colonial Legacy: Historians of Nature and Man's Nature,* edited by Lawrence Leder, 34–58. New York: Harper and Row, 1973.

"Indians as Ecologists and Other Environmental Themes in American History." *American Indian Environments: Ecological Issues in Native American History*, edited by Christopher Vecsey and Robert Venebales, 46–64. Syracuse, N.Y.: Syracuse University Press, 1980.

"Robert Beverly: Colonial Ecologist and Indian Lover." *Essays in Early Virginia Literature Honoring Richard Beal Davis,* edited by J. A. Lemay, 91–99. New York: Ben Franklin Co., 1977.

"Indian-White Relations, The Great Sioux Nation, the Treaty of 1868, and the Great White Father" [Wilbur R. Jacobs's testimony from the Wounded Knee Trial]. *The Great Sioux Nation Sitting Judgment on America,* edited by Roxanne Dunbar Ortiz, 60–61, 79–88, 116–18, 150. Berkeley, Calif.: Moon Books, 1977.

"British Indian Policies to 1783." *History of Indian-White Relations.* Vol. 4 of *Handbook of North American Indians,* edited by Wilcomb E. Washburn, 5–12. Washington, D.C., 1988.

"Willa Cather and Francis Parkman: Novelistic Portrayals of New France." *Willa Cather, Family, Community, and History,* edited by J. J. Murphy, 253–64. Provo, Utah: Brigham Young University Press, 1990.

ARTICLES IN REFERENCE WORKS, NEWSPAPERS, AND MAGAZINES

"No Silver Spoon: Washington Was a Self-Made Man." *Los Angeles Times,* 22 February 1953.

"Lincoln Was the Spirit of Democracy." *Los Angeles Times,* 12 February 1954.

"Lincoln, the Frontier Politican." *Los Angeles Times*, 11 February 1955.

"The American Past: Translating the Heretofore." *The Saturday Review* 44 (6 May 1961), 23–24.

"The 'Aged' Office of Governor of California." *Westlake Magazine* 3 (Spring 1969), 2–7.

"Frontiersmen, Fur Traders, and Other Varmints, An Ecological Appraisal of the Frontier in American History." *The AHA Newsletter* 3 (November 1970), 10–15.

"Lo the Poor Indian." *The AHA Newsletter* 4 (March 1971), 38–40.

"Roots of Wounded Knee." *New York Times*, 22 March 1973.

"William Hickling Prescott" and "Francis Parkman." *Encyclopedia Britannica,* 15th ed., 992–994, 1019–1020. Chicago, 1974.

ARTICLES IN ACADEMIC PUBLICATIONS

"Presents to Indians along the French Frontiers in the Old Northwest, 1748–1763." *Indiana Magazine of History* 14 (September 1948), 245–256.

"Wampum, the Protocol of Indian Diplomacy." *William and Mary Quarterly* 33 (October 1949), 596–604.

"Presents to the Indians as a Factor in the Conspiracy of Pontiac." *Michigan History* 33 (December 1949), 314–322.

"Was the Pontiac Uprising a Conspiracy?" *Ohio Archeological and Historical Society Quarterly* 59 (January 1950), 26–37.

"The Indian Frontier of 1763." *Western Pennsylvania Historical Magazine* 34 (September 1951), 185–198.

"Edmond Atkin's Plan for Imperial Indian Control." *Journal of Southern History* 9 (August 1953), 135–148.

"A Message to Fort William Henry: An Incident in the French and Indian War." *Huntington Library Quarterly* 6 (August 1953), 371–381.

editor and intro., "'Turner As I Remember Him,'—Herbert E. Bolton." *Mid-America—An Historical Review* 33, new series (January 1954), 54–61.

"Frederick Jackson Turner—Master Teacher." *The Pacific Historical Review* 33 (February 1954), 49–58.

"Wilson's First Battle at Princeton: The Chair for Turner." *Harvard Library Bulletin* 7 (Winter 1954), 74–87.

"Some Social Ideas of Francis Parkman." *The American Quarterly* 9 (Winter 1957), 387–96.

"Highlights of Francis Parkman's Formative Period." *The Pacific Historical Review* 27 (May 1958), 149–158.

"Some of Parkman's Literary Devices." *New England Quarterly* 31 (June 1958), 244–252.

"The Frederick Jackson Turner Papers at the Huntington Library," written in collaboration with Ray A. Billington. *Arizona and the West* 2 (Spring 1960), 73–77.

"Francis Parkman's Oration 'Romance in America.'" *American Historical Review* 68 (April 1963), 692–698.

"Frederick Jackson Turner—'The Significance of the Frontier in American History.'" *The American West* 1 (Winter 1964),25–32, 78.

"History and Propaganda: The Soviet Image of the American Past," with Edmond Masson. *Mid-America—An Historical Magazine* 16 (April 1964), 79–91.

"Frederick Jackson Turner's Notes on the Westward Movement, California, and the Far West." *The Southern California Quarterly* 46 (June 1964), 161–169.

"Research in Agricultural History: Frederick Jackson Turner's View in 1922." *Agricultural History* 42 (January 1968), 15–22.

"Turner's Methodology, Multiple Working Hypotheses or Ruling Theory." *Journal of American History* 54 (March 1968), 853–863.

"Wider Frontiers—Questions of War and Conflict in American History: The Strange Solution of Frederick Jackson Turner." *California Historical Society Quarterly* 42 (September 1968), 219–236.

"Frederick Jackson Turner's Views on International Politics, War, and Peace." *Australian National University Historical Journal* 9 (November 1969), 10–15.

"The Many-Sided Frederick Jackson Turner." *Western Historical Quarterly* 1 (October 1970), 363–372.

"Colonial Origins of the United States: The Turnerian View." *Pacific Historical Review* 40 (February 1971), 21–37.

"In Memory of Allan Nevins." *Pacific Historical Review* 40 (May 1971), 253–56.

"The Fatal Confrontation: Early Native-White Relations on the Frontiers of Australia, New Guinea, and America—A Comparative Study." *Pacific Historical Review* 40 (August 1971), 309–324.

"The Indian the Frontier in American History—A Need for Revision." *Western Historical Quarterly* 4 (January 1973), 43–56.

"The Tip of an Iceberg: Pre-Columbian Indian Demography and Some Implications for Revisionism." *William and Mary Quarterly* 31 (January 1974), 123–132.

"Native American History: How It Illuminates the Past." *American Historical Review* 80 (June 1975), 595–609.

"The Great Despoliation: Environmental Themes in American Frontier History." *Pacific Historical Review* 47 (February 1978), 1–26.

"Sherburne F. Cook, Rebel Revisionist." *Pacific Historical Review* 54 (May 1985), 191–199.

"National Frontiers, Great World Frontiers, and the Shadow of Frederick Jackson Turner." *International History Review* 7 (May 1985), 261–270.

"Columbus, Indians, and the Black Legend Hocus Pocus." *American Indian Culture and Research Journal* 17 (1993), 175–197.

"Francis Parkman and Frederick Jackson Turner Remembered." *Proceedings* of the Massachusetts Historical Society 105 (Boston 1995), 39–58.

INDEX